Environment and the Global Arena

Global Issues Series

Series Editors: James E. Harf and B. Thomas Trout
Volume Editors: James E. Harf and Kenneth A. Dahlberg

Environment and the Global Arena

Actors, Values, Policies, and Futures

Kenneth A. Dahlberg, Marvin S. Soroos,
Anne Thompson Feraru, James E. Harf,
and B. Thomas Trout

Duke University Press Durham, 1985

Printed in the United States of America
Library of Congress Cataloging in Publication Data
Main entry under title:
Environment and the global arena.
(Duke Press global issues series)
Bibliography: p.
Includes index.
1. Environmental policy—Addresses, essays, lectures.
2. Environmental protection—Addresses, essays,
lectures. I. Dahlberg, Kenneth A. II. Series.
HC79.E5E743 1985 363.7 84–21251
ISBN 0-8223-0621-2 (pbk.)

Contents

Figures and Tables

Figures

Tables

Series Preface

This text is one in a series of volumes on contemporary issues in the global environment. The Global Issues Series, of which it is a part, is the result of a multi-year project funded by the Exxon Education Foundation to develop educational resources for a number of problems arising from the shifting nature and growing interdependence of that environment. The issue areas addressed in this ongoing project include food, energy, population, environment, economic interdependence, development, arms and security, and human rights.

Each of these issues has been addressed within a systematic and integrated framework common to all. After establishing the substantive dimensions—such as the historical evolution, the structure of its global system, its basic contemporary characteristics—needed to provide the requisite foundation for inquiry, this framework is applied in separate chapters pursuing four distinct analytical perspectives: (1) Who are the global *actors* involved in the issue and what are the linkages among them? (2) What prevailing *values* are operating and how have the relevant actors responded to these values? (3) What *policies* are applied by these actors at the global level and how are these policies determined? and (4) What *futures* are represented in the values and policies of these global actors? The relationship among these perspectives and their use to link analysis of the various issues are illustrated in figure 1. Each segment—actors, values, policies, and futures—then represents a distinct analytical approach.

In addition, differentiating this project and its product from other texts, each of the single issue volumes incorporates exercises that afford the student the opportunity to engage in a variety of active learning sequences—should the instructor so desire—in order to

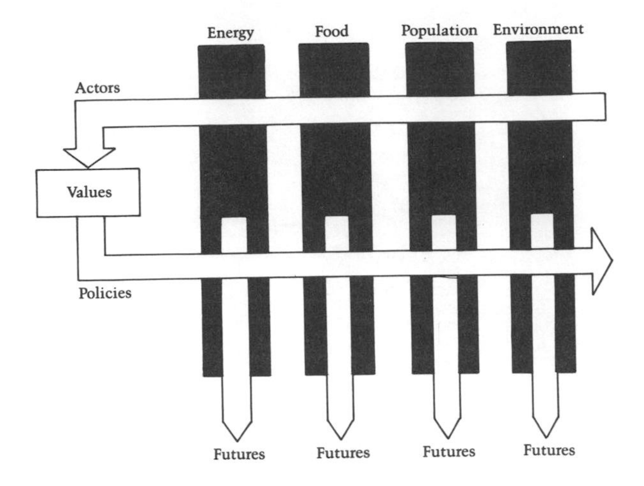

Figure 1 Actors, values, policies, futures.

understand better the complexities of the issue.

In realizing its goals, the project enlisted the participation of thirty-one prominent scholars who worked together with the project directors. These individuals contributed their substantial expertise both to specific issue areas and to the analytical development of the project as a whole.

Each volume of the series represents the specific contributions of three principal authors. One, the issues specialist, was responsible for the substantive introduction and, in some cases, the summary conclusion and for supervising and coordinating the efforts of the other two authors. Those authors applied the components of the perspective framework—actors, values, policies, futures—to the issue in separate chapters. The series editors assisted in the writing of these chapters, then edited each individual volume together with the relevant issue specialist. The editors incorporated the feedback from pedagogical and substantive specialists as well as from field testers.

The project evolved through a series of stages centered on two development workshops—the first for the "issue" and "perspective" specialists, and the second for all participants—and the elaborate field-test network of the Consortium for International Studies Education. The materials developed in the workshops were produced in field-test editions which were then used across the country in a variety of instructional settings. All the volumes have therefore been revised and refined based on actual classroom experience.

We wish to acknowledge formally the financial support of the Exxon Education Foundation, which provided the means to develop this project, and we wish to express our gratitude to several individuals: Roy E. Licklider, Rutgers University, who while serving with the Exxon Education Foundation encouraged us initially to pursue this venture; Richard Johnson who as our contact at the Exxon Education Foundation shepherded this project to completion; and Suzanna Easton of the Department of Education for her enthusiastic support and encouragement not only of this project but of the entire educational enterprise that it represents.

Two books from the Global Issues Project have been published by Holt, Rinehart and Winston: *Population in the Global Arena* and *Food in the Global Arena*. Duke University Press has published three texts in the series: *Energy in the Global Arena, Environment in the Global Arena,* and *The New Global Agenda*. We are grateful to a number of individuals who encouraged us at various stages of the publishing process. First we worked with Patrick Powers who as a senior editor at Holt believed in this project. Marie A. Schappert of Holt brought a sound, businesslike style to her task of overseeing the first two manuscripts. We are especially indebted to Richard C. Rowson, director of Duke University Press, who has continuously demonstrated a strong commitment to this project. Reynolds Smith, as editor responsible for the three Duke University Press manuscripts, translated them from final draft into print with a very high level of expertise and sound professional judgment. Both Mary Mendell and Bob Mirandon at Duke Press also contributed materially in the quality of the manuscript. We are also grateful for the help of our administrative assistant and typist, Edith Bivona, who delivered work of the highest quality under less than ideal circumstances. She brings a pleasant manner and much enthusiasm to her work, making our professional lives much easier. Finally, we would like to acknolwedge the support of the Ohio State University's Mershon Center and its director, Charles F. Hermann, who was forthcoming with assistance when needed.

James E. Harf
B. Thomas Trout

Preface

The environment has taken on new meaning as we approach the eve of the twenty-first century. For scientists, the environment has always been viewed as a composite of natural and physical structures exhibiting a complex variety of linkages, although even they did not have a clear understanding of the extent of these relationships until recently. Now average citizens, who once thought of the environment as the "outdoors," that part of their surroundings to be used in the pursuit of fulfillment and enjoyment, are becoming aware that the "environment" is something more than wilderness areas and streams. Rather a significant number of these citizens have also begun to view their total surroundings as a complex ecosystem with a wide range of interrelated plant and animal species existing within a diverse and comprehensive global environment.

It is not surprising that there is now a heightened awareness on the part of both scientists and observers. Environmental research has consistently yielded evidence to suggest that linkages throughout the system are both more pronounced and more intricate than originally thought. Moreover, the capacity of the human race to affect this environment—for good or ill—in ways and at levels unforeseen one hundred years ago has demonstrated the scope, intricacy, and fragility of these relationships.

In 1972, the United Nations Conference on the Human Environment was held in Stockholm, Sweden. Though typically thought of as the beginning of a concerted global effort to deal with environmental issues, in reality the Stockholm Conference represented the culmination of efforts in the post–World War II era to bring together key actors of the globe to address the issue. The conference was, in the words of observers, "the successor to the great voyages of discovery and exploration that made people aware of the shape of the world and the diversity of its lands and waters, rocks, vegetation, faunas and cultures."[1] International cooperation during the 1957–58 International Geophysical Year (IGY) demonstrated that coordinated effort could work and, moreover, that it was essential if the global community were to acquire the necessary information about the far corners of the earth. The IGY, in turn, spawned a number of joint ventures to unlock the secrets of the environment, including the Upper Mantle Project (1964–70), which examined the lithosphere (the outer portion of the surface of the globe, i.e., its mantle); the Global Atmospheric Research Program (1970–80); the various projects of SCOPE (Scientific Committee on Problems of the Environment sponsored by the International Council of Scientific Unions); the UNESCO program on Man and the Biosphere, and programs of the United Nations' Food and Agricultural Organization (FAO); the World Health Organization (WHO) the World Meteorological Organization (WMO); and the International Union for the Conservation of Nature and Natural Resources (IUCN).

The Stockholm Conference was the culmination of a number of major developments.[2] First, the scientific community was joined by active nature protection elements under the aegis of professional ecologists. Second, the concern for the environment spread beyond the developed Western industrialized nations to both the socialist and developing worlds. Finally, the approach changed to embrace a much broader conception of the term environment. All aspects of the natural environment became matters of concern: "Land, water, minerals, all living organisms and life processes, the atmosphere and climate, the polar icecaps and remote ocean deeps, . . . space, . . . the human situation, . . . and the relationship between man-made and natural environments."[3] Perhaps the most important manifestation of this commitment to the environment is the official body created for implementing the precepts of the Stockholm Conference, the United Nations Environment Program (UNEP). This mechanism represented the synthesis of ideas for action that emerged at Stockholm. It became, in short, a coordinating body, which operated under the fundamental principle that responsibility for addressing environmental issues rested not with a specialized body but with every actor. UNEP, in turn, would become a creative body, an idea generator, a coordinator, and an evaluator. This strategy takes cognizance of the global dimensions of environmental issues.

Let us begin then to understand why it is that we consider environment as a *global issue*. Indeed, what makes an issue global? We must think in terms of the globe itself. Global issues by definition *transcend the traditional boundaries of the nation-state* or the regional conjunction of nation-states. We are therefore addressing issues, like environment, whose impact will be felt beyond a clear confinement in terms of political or even geographic space. Such issues necessarily affect the judgment and actions of large segments of the world's inhabitants either directly or indirectly. Recognition of that fact, however, simply begs the question again: What are the characteristics of environment that define its impact beyond such previously recognized and accepted limits?

First, such issues are characterized by an *incapacity for autonomous decision*. No single actor, or corporate group of actors, is capable of resolving the persistent problems associated with the global environment—climatic change; atmospheric and water pollution; the loss of cropland, rangeland, and forests; species and habitat loss; the potential effects of nuclear and conventional war; and increasing resource limits. Because environmental problems know no national boundaries, resolution of such issues requires (and will continue to require) the cooperation of a wide range of actors, including not only nation-states throughout the world, but also a host of others, governmental and nongovernmental, which operate at the international, national and subnational levels. While on the one hand national governments have stepped up their efforts at addressing environmental ills (witness the proliferation of environmental agencies within the United States Government), there has been on the other hand a dramatic increase in both the number and activity of institutions organized from the grass roots to the international level. The former group of organizations has emerged primarily as a consequence of frustrations expressed by citizens as they perceive, rightly or wrongly, that national governments are unwilling and/or unable to address environmental concerns successfully. The latter type of institutions arose because nation-state governments themselves recognized that cooperation across countries could be expedited and enhanced if formal mechanisms were created to deal with environmental issues on a global basis. Earlier we chronicled the nature of activity at the global level. Structures were created for the purpose of trying to understand better the parameters of some environmental problem, or of seeking transnational cooperative ventures to address one or another concern. UNEP represents the pinnacle of such activity. Simultaneously, international organizations designed for other purposes have undertaken an environmental role as part of their agendas, approaching environmental concerns from the perspective of their own areas of responsibility. Thus, for example, FAO undertakes forest and soil management practices, while WMO monitors soil erosion through the use of satellites. Nongovernmental groups of a transnational nature were well represented at Stockholm as some 237 such organizations representing 21 different fields of interest played vital roles. Three particularly influential organizations were the International Council of Scientific Unions, the International Union for the Conservation of Nature and Natural Resources, and the Friends of the Earth. The *actors' perspective* is therefore an important component of a comprehensive strategy for addressing environment as a global issue.

A second characteristic of global issues is that each possesses a *present imperative*, which not only impels various actors to press for resolution, but which encompasses the varied and often competing views as to how that resolution ought to proceed. Most actors agree upon four values growing out of the major environmental problems discussed earlier: controlling pollution, preserving genetic diversity, conserving natural resources, and limiting population growth. Support for controlling pollution in the air, in water, or on land came early in the environmental movement and has developed a broad base among wide segments of the global society. The reasons are obvious. Much (but not all) pollution is readily observable, easily documented, and a clearly defined nuisance. The Industrial Revolution brought with it a capacity for widespread pollution and, in the absence of effective concern, the human race, concentrated in the urban centers of developed societies, learned to live for over two hundred years with its consequences. In the post–World War II era, as population ballooned in the developing world, these societies increasingly suffered from their inability to prevent pollution as a normal by-product of living. Actors also seek resolution of the remaining three values, although with respect to limiting population, a small number of countries believe that their interests are best served by encouraging growth. It would be a mistake to conclude that because in the abstract most actors support these goals, there is broad agreement over the relative importance of each value and the extent to which each should prevail when it comes up against a nonenvironmental value (such as the need for more food or energy). There are wide ranges of importance assigned to various ecological issues by actors as they weigh the need for concern about the environment against the opportunity costs of such concern. Moreover, it would also be misleading to conclude that, even where there is wide agreement that an undesirable condition exists, there is also consensus on how the global community ought to respond to the problem. Hence, a *values perspective* provides a second analytical focus in this volume.

A third characteristic of global issues, not peculiar to such issues but nonetheless integral to the sequence already defined by actors and values, is that their *resolution requires policy action.* It is evident that action—the process that combines actors with values—implies policy, whether it is formal policy articulated by governments and organizations, or relatively undefined policy that guides individuals in their everyday life. Environmental policies tend to adopt several different approaches to the problem. At the one end of a policy continuum, decision makers urge voluntary restraint, although its success depends on widespread, even universal agreement on the costs of failure to comply, a highly unusual situation. For example, the Limited Nuclear Test Ban Treaty of 1963—which banned the testing of nuclear weapons in the atmosphere, in space, under the ocean, and even underground if the damage spilled over to a neighboring country—underscores this point. This treaty reflects a kind of voluntary restraint in that nations choose to subscribe or not to subscribe to its provisions and thus make their own decision about compliance. Even though the environmentally related damage level of such blasts is well documented, many nations, most notably France and China, have not signed the treaty and refuse to abide by it. At the other end of the continuum is the notion of public (i.e., global) responsibility for the resources, actions, and consequences that constitute the environmental problem. This is difficult to achieve because there is no global public enterprise that would assume and enforce the provisions of public ownership even if this strategy were a desirable one. The recent attempts to establish an International Seabed Authority to control the resources of the ocean floor demonstrate that, although possible, even a relatively minor effort at public ownership is not without major difficulty. Between the extremes of voluntary restraint and public ownership lie a number of general types of policy options. All require that global actors be bound by some prescription, whether the strategy takes the form of enforced behavior, regulations or other legal restraints, enforced contribution or support (such as taxation), complete liability for actions affecting the environment for identifiable transgressors, or some other form. Without the intervention of comprehensive policy action, it is unlikely that these problems will be resolved. The *policies perspec-*

tive is therefore critical for understanding the interrelatedness of issues and actors affecting environment in the global arena.

A final element that characterizes global issues is their *persistence.* These environmental concerns draw us to examine the future, to look to individuals, groups, and institutions who make it their business to forecast what lies ahead. Although humans have adapted more or less to their current environment, the future is of far greater concern because of current observable trends. However, prognosticators disagree about the prospects for and the magnitude of increasingly adverse conditions in the future. Both individual analyses and major collaborative reports on the future display this lack of consensus. All do agree that the problems must be addressed squarely in a comprehensive analytical manner if the human race is to have any chance of surviving on this increasingly fragile planet. The human race has demonstrated its capacity to use science and technology to achieve vast magnitudes of change for both good and ill. But future strategies require not only that we use science to create physical opportunities for change; but also that we reevaluate political, social, and economic systems and processes of the past and present in order to determine how alternative structures outside the technical area can aid in effecting a better environment. Six different strategies, discussed in this text, are typically advanced to achieve this goal: (1) the creation of new international regimes; (2) reliance on steady-state systems; (3) centralized political authority; (4) a return to more manageable institutional arrangements ("small is beautiful"); (5) a concerted effort to redistribute wealth and resources from the "haves" to the "have-nots," whether individuals or nations; and (6) a complete reliance on technology (the "technological fix"). The trends and consequences of forecasts and alternatives for global environment are addressed in the final perspective of this volume, the *future perspective.* In the concluding chapter we step back and reflect once again on progress recently made and draw hope for the future.

Several individuals played critical roles in this book. Kenneth A. Dahlberg, together with the series editors, supervised it from conception to completion—selecting the other principal authors, adapting the general framework to the environment issue, evaluating the reviews, and editing the field-test edition and final manuscript. Dahlberg also contributed the first chapter, which describes the background and global characteristics of the environment problem. Marvin S. Soroos wrote the chapters that focus on the *values, policies,* and the *futures* perspectives, and assisted in editing the entire manuscript. Anne Thompson Feraru wrote the chapter on the *actors'* perspective. The series editors wrote both prefaces and the concluding chapter. In addition, a number of persons critiqued this volume in draft form, and contributed valuable suggestions, and some used it in their classrooms. The material was also used in a series of workshops for faculty across the country, and many insights and amendments were offered. Finally, we wish to acknowledge Robert S. Jordan, Chadwick F. Alger, Richard W. Mansbach, and Dennis Pirages who examined, respectively, the actors, values, policies, and futures chapters.

James E. Harf
B. Thomas Trout
September 1984

1 Environment as a Global Issue

What Is the Environment?

Most Americans tend to understand "the environment" as something separate from their everyday lives, a place that they may go for recreation: picturesque wilderness areas, clear flowing mountain streams, exotic big game in national parks, distant steaming tropical jungles, and so forth. This view (which is shared by the Europeans) is in sharp contrast to the way in which many members of traditional societies understand the environment. To them, the environment is a complex web of relationships and interdependencies of which they are a part. They exist within their surroundings, feeling and perceiving, influencing and being influenced by them. They are a part of their environment and their environment is part of them; they have a sense of mutual interdependence for long-term survival. Modern industrial man has lost this traditional sense of environment and has only begun to appreciate the similar but more sophisticated understandings of the ecological and environmental sciences.

Public awareness has been helped by satellite photographs of the earth that have for the first time given us a visual sense of our global home, a realization that in spite of its vastness as seen from our individual locales, our delicate and beautiful blue planet is a self-contained unit floating in space. Phrases such as "ecosphere" and "spaceship earth" have taken on a new meaning as a result. Also, people can more easily appreciate the interrelatedness of the earth's basic land, water, and atmospheric systems. The oceans and seas interconnect with currents flowing gracefully around the continents producing climatic, weather, and cloud patterns that move across the large terrestrial features of the earth's surface: mountains, deserts, forests, canyons, lakes, and so on. One can see more clearly than ever before that pollution of air or water systems becomes part of these great flows without reference to the artificially drawn lines of national boundaries.

This contrast between the commonly shared natural resources of the globe and the attempts of various institutionalized groups to control, exploit, or manage those resources is one of the major themes of this text. There are shared threats as well as shared resources. Here, too, various groups seek to minimize those threats to themselves without regard to how their actions may increase the risks or damage to others. Resources are ultimately limited both in their availability and their useful quality. We have become most aware of this recently in the area of energy, where we are currently dependent on fossil fuels, which are finite. As the easily obtainable sources are used, it is necessary to expend more and more energy to extract what remains, which is found in more remote places (like the North Sea or Alaska) or which is of lower grade (shale oil or tar sands). In a similar way our biological resources are also limited as are the great reservoirs of air and water.

Key Concepts

Ecologists and environmentalists have developed several concepts that can help us to understand the environment better. The three major concepts are *ecosystem*, *carrying capacity*, and the *tragedy of the commons*.

Ecosystem

The first concept, ecosystem, in nontechnical terms can be said to stress the interrelatedness of a wide range of species living in common in a given environment. Each species has a number of relationships with other species (both plant and animal) as well as with the energy, nutrient, and related physical elements of the system (see figure 1.1). Ecologists stress at least two important points about ecosystems. First, changes in one part of the system may have profound effects on other parts of the system, effects that can be understood and estimated only if one analyzes the system as a whole and takes the time to know its various interrelationships. Second, the more *diverse* an ecosystem is (that is, the wider the variety of distinct species living in complex webs of relationships), then generally the more stable, resistant to disruption, and adaptable it will be. Thus, for example, traditional forms of agriculture that use a

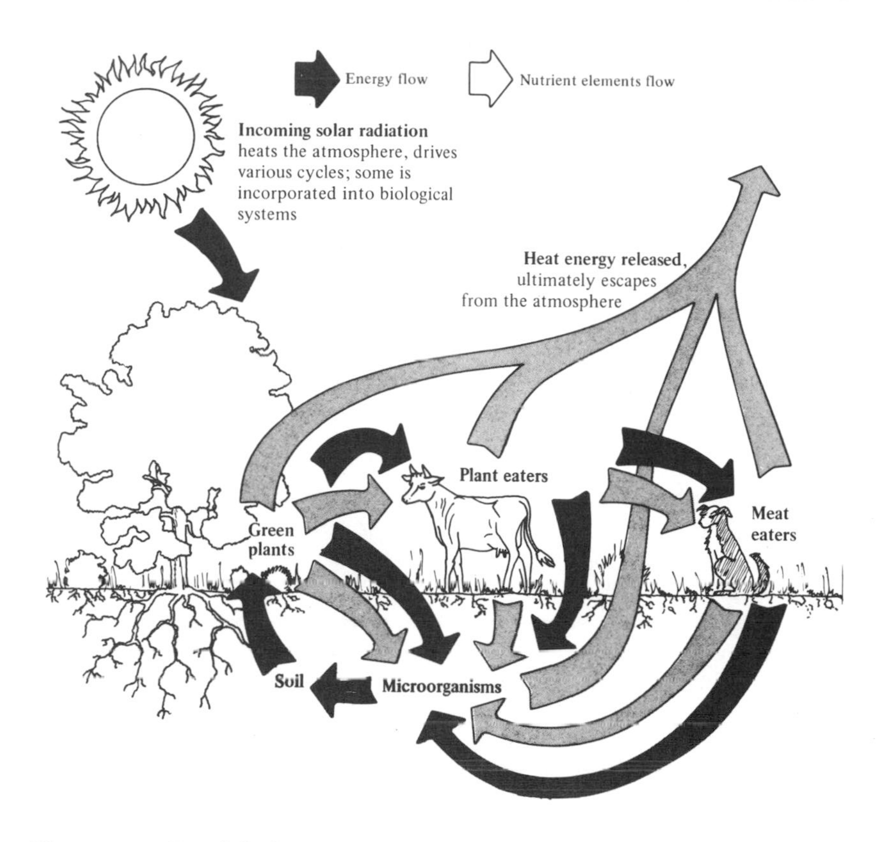

Figure 1.1 Simplified representation of an ecosystem. Source: N. J. Greenwood and J. M. B. Edwards, *Human Environments and Natural Systems*, 2d ed. (North Scituate, Mass.: Duxbury Press, 1979), p. 17.

number of varieties in the same field—some resistant to drought, some to frost, some to moisture and blight, others to various pests—are much more resistant to climate variations and pest outbreaks than are forms growing only a single variety hybrid. Extensive, industrially supported agriculture tries to compensate for such built-in instability by external and technological means: the use of chemical pesticides and herbicides, careful irrigation, etc. Even so, there have been serious and dangerous outbreaks of crop diseases, such as the widespread corn blight in the United States in 1970, which caused losses of up to 50 percent of the crop in many southern states and 15 percent nationwide.

Carrying Capacity

A second concept is somewhat more related to managed ecosystems than wild ones. In ecological applications, carrying capacity refers to the maximum rate of exploitation of given biological resources which can be carried on indefinitely. A common example is cattle ranching. The number of livestock that can be grazed indefinitely on a particular rangeland will depend upon a number of interrelated factors. It will depend of course on the size of the herd, its general health, and its reproductive rate. Also, it will depend upon the health of the supporting ecosystem—the availability of grass and water especially. The latter are affected by the number of cattle (too many cattle leading to overgrazing and the pollution or depletion of water) as well as by weather conditions, pests, and diseases. Finally, the carrying capacity of the environment will depend on the size and demand of the animals themselves. If a new breed of cattle is introduced that weighs half again as much, eats twice as much grass per day, and drinks twice as much water per day, the rangeland will be able to sustain far fewer animals. Similar factors must be considered in addressing the carrying capacity of the earth in sustaining human existence. For example, the average American uses anywhere from five to twenty times as much in the way of food, water, mineral, and energy resources as the average person in India. Thus, the carrying capacity of the earth is directly linked to the demand or consumption patterns of its inhabitants, not just to the numbers.

Tragedy of the Commons

A third concept, tragedy of the commons, is named after a much-heralded article by ecologist Garrett Hardin.[1] He analyzes early English grazing practices to illustrate the dilemmas involved in the management of shared resources. The village "commons" was the open grazing land shared by villagers. Because each villager had little incentive to restrict the number of cattle grazed (since others would increase their herds even if he did not), the commons often became severely overgrazed and everyone suffered (Hardin and others suggest various ways to try to mitigate tragedy-of-the-commons type situations, which are discussed in detail in chapter 4). The important point here is that much like individual villagers, various states and multinational corporations of the world can often be seen to act in the same way regarding the "global commons."

They often overfish the oceans, reducing permanently the catch of valuable species. They pollute the atmosphere without adequately considering the material and health costs to their own and other countries. They use the oceans as a dumping ground without knowing what the losses to fisheries, human health, and basic oxygen production may be. They cut down massive forests without regard for the loss of capacity for regeneration or the increases in soil erosion. Nor do they consider how such deforestation may affect both local and global rainfall and temperature patterns.

There are, moreover, a number of added dimensions to the tragedy of the commons at the global level as compared to the village level. First, the different geographic locations and resource endowments between countries inevitably mean that any particular activity will have very different impacts, costs, and risks from country to country. For example, in recent years, the French have been dumping large amounts of waste salts from their mining operations into the Rhine River. This is an easy and inexpensive way of disposing of this waste (more properly, this action is dispersal rather than disposal of waste). However, there is a high cost downstream to the Dutch, where the levels of salt have made the Rhine unusable as a source for drinking water and created serious problems for its use in irrigation for agriculture. The result has been a serious controversy between the French and Dutch governments over how this common resource is used. Many similar examples can be found for wastes disposed of in the atmosphere.

A second factor that distinguishes the global commons relates to the different capacities that countries have to exploit various resources. Unlike the farmers who had very similar "grass harvesters and converters" (cattle), some countries have extensive capacities and technologies to exploit resources such as oil, fish, manganese nodules off the bottom of the ocean, etc., while others have little or no capacity. The rough division of the world in this regard is between the industrial countries of the northern hemisphere and the poorer, less industrialized countries of the southern hemisphere. Thus, the question regarding the sharing of the common resources of the international parts of the oceans has become much more than each country having equal access to these international waters and to the fish, oil reserves, and minerals found there. In the negotiations on these issues that were carried on in the Law of the Sea Conference recently, the question became one of how the countries without high levels of technology are still to be assured that they receive a fair share of what is actually removed from these global commons by the technologically advanced countries and corporations.

A third factor, related to the first two, is that each country has its own set of national interests, which derive from its history, its location, its resources, and its economic and technological capabilities. For example, Iceland and the United Kingdom approach coastal fisheries quite differently. Both have good fishing waters off their coasts. On the one hand, the U.K. has traditionally defended the idea of only limited national jurisdiction and control over coastal waters, preferring the long-established three-mile limit for sovereign control. This approach was preferred because the U.K. had a relatively small fishing industry as compared to her other industries, and because with her large navy and extensive trading, her fishermen could benefit by fishing off the coasts of other countries, even if others fished close to her shores. Her fishing fleet was also more advanced technologically, so British fishermen tended to gain overall. On the other hand, fishing is by far the largest single industry in Iceland, involving thousands of small boats. Because of this, Iceland sought protection from the much larger and more advanced boats of the British, the Danes, and the Norwegians by claiming a fifty-mile territorial limit around her coast. Iceland's attempt to exclude others from fishing in these waters led to the so-called cod war between her and the British.

Our discussion has moved from a general description of globally shared resources to examples of how different nations come into conflict over the use of these resources. A number of basic points about environment as a global issue have been suggested in this discussion. First, from a global perspective, there are a number of shared resources or commons, which are both the heritage and responsibility of mankind. Second, these resources are limited either in their availability or in their productive use. Third, countries and other organized groups, such as multinational corporations, have their own particular interests, which often do not coincide with the larger global interests of all of humanity. If these conflicting or partial interests cannot be moderated, mediated, or expanded to fit the larger global interests, then we may well face the prospect of a global tragedy of the commons.

Cultural and Historical Roots

Our current environmental problems have developed over the centuries and have deep cultural and technological roots. In this section we will review the main cultural differences between Western and non-Western peoples and then turn to an examination of the gradual evolution of current threats to the biosphere.

Different Cultural Views of Nature

It is difficult for Americans and Europeans to realize that their understandings of nature, the environment, and society are culturally bound. And since cultural differences in the conception of nature and man's relationship to it relate historically to rather distinct behavior patterns in regard to the environment, it is all the more important to explore them. Let us turn first to the Judeo-Christian belief that man is separate from, and above, nature and the beasts, and should hold dominion over them. Although there is an implied idea of stewardship involved here (man should respect nature as part of God's creation), the themes of hierarchy (where man is placed above the beasts but below the angels) and dominion (in which nature is a possession of man) have tended to be the dominant ones. That is, while seeking God's purposes, man should use and enjoy the bounties of nature.

The early Christian beliefs of the separation of man from nature and man's free use of nature were, of course, placed in a larger framework of man pursuing God's will and seeking salvation thereby. This striving for an ultimate heavenly existence provided a spiritual concept of progress, the idea of moving through time toward a place of eventual perfection (or damnation). During the tumultuous transformation that occurred as the feudal system and the Holy Roman Empire broke down, this spiritual idea of progress was transformed into a secular version of progress focused on the quest for perfection in man's material existence rather than in his religious existence. This transformation involved a complete restructuring of both the intellectual and the social world. Not only was the idea of progress made secular, but the whole notion that societies and their rulers were subordinate to God's will (as interpreted by the Catholic Church and the Pope) was altered. The Renaissance, the Reformation, and the rise of rationalism and science all combined to weaken the authority of the Church, especially over matters of governing society. When kings were no longer seen to have "divine right," the philosophical foundation for the ultimate authority of the ruler became a burning question. The question was first answered by the idea of a social contract between the ruler and the ruled, which led eventually to the idea of democracy. The Reformation challenged the authority of the single Catholic Church to speak in God's name: rather, each individual should be able to read God's words in his own native language. Thus, nationalism as well as democracy was given an early foundation upon which to build. Rationalism and science suggested that man could both understand and control nature, and to this emerging faith was added the new idea of progress, which suggested that man could achieve perfection on earth, rather than in the afterlife. Sir Thomas More's *Utopia* was an early embodiment of these ideas.

As suggested, this new idea of progress was primarily material, rather than spiritual. New disciplines gradually grew up to explain and facilitate the exploitation of nature, which was now seen as an object of human enterprise, rather than part of God's wondrous creation. Scientists sought to understand the various "secrets" of nature. Technologists, and later engineers, sought to devise tools and machines to extract what man needed from nature. Economists also began to conceive of nature as simply an object to be exploited. They postulated that air and water were 'free goods," which each person or corporation could use as they saw fit and without any economic calculation as to cost or responsibility for their purity or maintenance. Or to phrase it another way, environmental pollution was defined as an "externality," which economists chose not to try to analyze or to address. The power of these new intellectual and mechanical tools to exploit nature became clear in the Industrial Revolution when new cities and railroads were built while forests were ravaged and rivers polluted. Even so, a strict Protestant ethic, which stressed frugality, saving for the future, and minimal ostentation, provided some restraint upon rampant individualism and excessive use of resources. Today, in an age of advertising and consumerism this restraint has been considerably weakened.

It should be emphasized, however, that in spite of this transformation and the rise of industrial society, a number of subcultures and traditions that have stressed simplicity, harmony with nature, and conservation have persisted. These have now combined with a number of new events, trends, and ideas to form several vigorous movements that challenge the foundations of the modern state system and modern industrial society. Indeed, many are suggesting that we may be entering a new period of transition comparable to that from the Middle Ages to modern industrial society. The environmental challenges involved range from the threat of nuclear war, which if it were to occur would plunge the globe into environmental disaster, to the basic findings of modern environmental and ecological science that man is not and cannot be separated from nature and that he must come to understand himself as an integral part of the larger natural processes.

In briefly outlining some of the main beliefs and characteristics of Western culture as they apply to the environment we can see that they are dynamic and expansive. Before examining how they have had an impact upon other cultures and environments, let us first describe some of the broad attitudes of non-Western cultures toward the environment. There are, of course, a number of differences among the many non-Western cultures and religions; here we will examine only those areas where they tend to have similar basic views or attitudes.

One general tendency found in traditional non-Western cultures is to view man as an integral part of nature, often even a subordinate part. Nature is commonly seen in spiritual terms. It is not simply an object of man's enterprise but is filled with good and evil spirits and is tied directly to religious belief systems. Hence, man does not control nature, but he tries to work with it, using various rituals, symbolic acts, and religious practices to try to understand and fit into nature's whims, rages, and cycles.

Westerners often judge such practices from the perspective of their own beliefs, referring to non-Western ceremonies as "primitive." One source of such judgments is the tendency to assess the verbal description or the external characteristics of a particular ritual or practice, rather than its function. For example, in New Guinea, every so often several tribes slaughter the wild pigs of the forests and have a ritual feast offering up the pigs to their ancestors. The verbal reasons given for the feast—the honoring and placating of dead ancestors—appear to place it in the realm of a primitive belief or myth that has no scientific or other useful basis. However, when one looks in great detail at the actual operation of this ritual and its impact upon the surrounding ecosystems, one finds that the ritual not only protects the forests from serious overuse and soil erosion caused by the pigs but provides some critical nutritional requirements for the tribes.[2]

Differences in Western and non-Western thinking also flow from different beliefs about the processes of change, progress, and the ability to master nature. The Westerner believes that man can and should master nature and that one who does not is "fatalistic." The non-Western person certainly sees and believes in change, but sees it more in terms of the natural cycles of life, death, and rebirth that he observes in the forests and fields around him. Hence, the Western idea of progress is not shared. Changes may be good or bad, but there is no automatic way in which good will triumph. Finally, the non-Westerner recognizes that he is not all-powerful and that one must work within the limits of his knowledge, society, and environment. Each person understands himself or herself not as an isolated individual, but as part of an extended family with links to the tribe and the chief, and as part of an extended cycle, with links to the fields and the forests. Each person is part of a complicated web that is not fully understood, but that one respects and worships.

Such views and beliefs are typically found in the rural areas of the developing countries, although they are coming under increasing challenge from mass communication media, particularly the transistor radio. It is important to recognize that these traditional views are rarely held by the elites of these countries because they have themselves become largely Westernized. The goals of these elites usually include rapid modernization and industrialization, something that foreign aid programs, trade patterns, and multinational corporations all encourage. Being exposed primarily to these elites, Westerners often assume that they represent the basic views and beliefs of their peoples. The result is a lack of awareness that substantial numbers of people continue to hold cultural views that are very different from one's own.

These different cultural values often result in different long-term behavior toward the environment. Even so, there are other factors that can override the verbal or cultural respect that a society may have for the environment. The press of modernization often imposes cultural dilemmas. A striking example is China. Traditional Chinese culture venerates the environment, yet largely due to its increasing population as well as to the regular battles of regional warlords, events in China have inflicted serious deforestation, causing soil erosion and therefore floods in the countryside. Veneration of the environment has been and continues to be challenged by the Chinese drive toward modernization.

Having examined some of the major cultural differences regarding nature and man's place in it, we need to turn now to an overview of how man's activities over the past several centuries have increasingly come to threaten the biosphere. This is also a paradox. The Western industrialized nations, culturally inclined to master nature and consequently tending to abuse it, have nonetheless, because of the same cultural inclination, been the first to put environmental protection on the policy agenda and the first to take action to redress the damage to the environment.

Evolution of Global Threats to the Biosphere

Man has always had an impact upon his environment. The Romans turned large parts of North Africa into deserts. As mentioned above, the Chinese often suffered from floods caused by mistreatment of the land. However, both the degree and extent of modern industrial man's impact upon the environment are unprecedented. Before examining the evolution of that impact and its severity, let us first review some of the main characteristics of the biosphere. At the broadest level, the biosphere is that part of the planet where the great physical systems of the earth interact with a variety of complex biological systems. The physical systems derive from such forces as gravity; the rotation of the earth; the radiant energy of the sun; and their interactions with the oceanic, terrestrial, and atmospheric masses. The biological systems of the earth have evolved over millions of years and have come to weave themselves into these physical systems in a number of cycles. Scientists have tended to examine these cycles in terms of many of the basic elements that flow through them: water, oxygen, nitrogen, carbon, and so forth. Several of these are illustrated in figures 1.2 through 1.5. However, it is important to realize that these and other cycles are not only operating simultaneously (if at different rates), but that basic aspects of the cycles are evolving and thus the cycles themselves and their relative balance are being transformed over time. In addition, there are both natural and man-made disruptions that can influence these cycles. Large volcanic eruptions have been known to reduce the amount of sunlight reaching the earth because of the reflective nature of volcanic dust in the upper atmosphere. This in turn can bring on regional climate shifts, including extremely cold winters and cool summers.

Historically, man-made disruptions have involved the use of fire in hunting, the movement from one field to another for cultivation, and the ever-increasing elimination of forests to provide more agricultural land. Over the past several centuries human impacts on the biosphere have been increasing at almost geometric rates as the two great science-based explosions in population and industry have spread around the earth. This spread has been uneven in several important ways. The population explosion has been the result, in

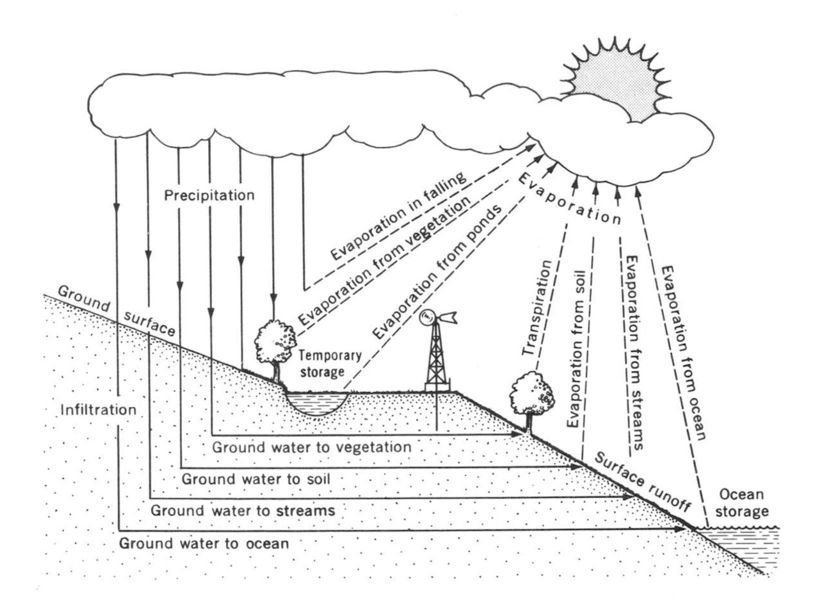

Figure 1.2 The hydrologic cycle. Source: Peter J. Richerson and James McEvoy III, *Human Ecology: An Environmental Approach* (North Scituate, Mass.: Duxbury Press, 1976), p. 29.

part, of well-intended but narrowly conceived public health programs. In the industrial countries, the impact of public health programs and other developments such as refrigeration, canning, pasteurization, and so forth was piecemeal and slow and went along with improving living standards throughout society. Although there was a population explosion in the industrial countries, many of the pressures were relieved through the opening of colonies, first in the New World, and then later in Asia and Africa. Moreover, this explosion was built upon a much smaller base and occurred over a much longer time frame. When public health measures were introduced into the developing countries, especially after World War II, this was done on a massive scale and involved sanitation, immunization programs, and the extensive World Health Organization campaign against malaria. Death rates, especially infant mortality rates, were brought down dramatically. However, birth rates remained high for a number of reasons. First, people want children. Second, children are an important source of labor for farms in many developing countries. Third, in many parts of the world, children traditionally have been the only source of support for couples once they reach an

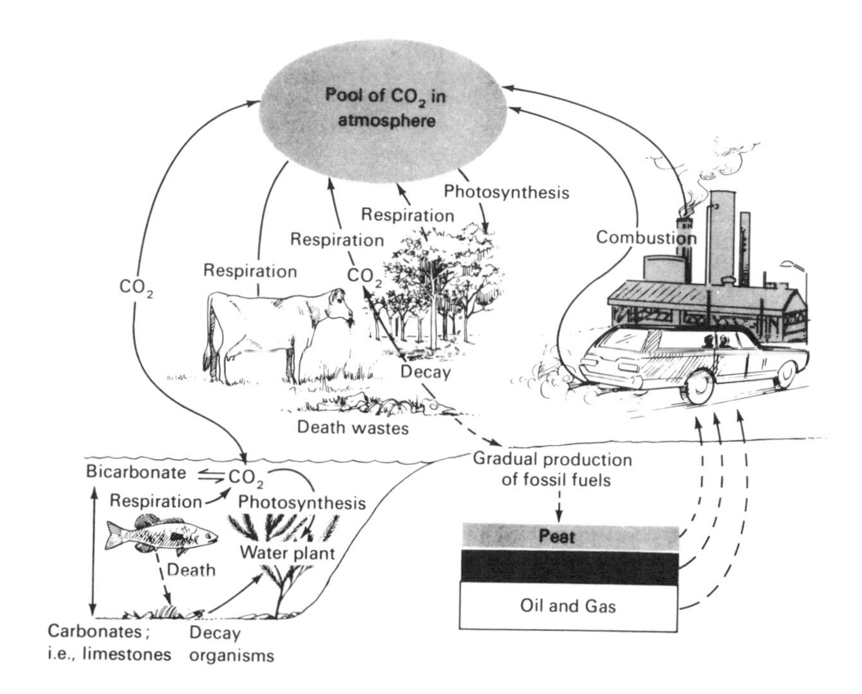

Figure 1.3 The carbon cycle. Source: Paul R. Ehrlich, Ann H. Ehrlich, and J. P. Holdren, *Human Ecology: Problems and Solutions* (San Francisco: W. H. Freeman, 1973), p. 152.

age when they can no longer support themselves; they are, so to speak, the couple's "social security," when no other is provided. Fourth, while public health interventions did not encounter serious cultural or technological barriers, family planning, population, and contraceptive programs regularly encounter a wide range of social, religious, and technical barriers to their widespread acceptance.

The spread of industrial society has been equally uneven. Part of the reason for this is cultural. As suggested earlier, Western culture was much more receptive to the systematic development of science and technology. Another basic reason relates to climate. It is not by accident that the industrial countries of the world today are almost exclusively located in the temperate zones (mostly in the northern hemisphere, but with a few examples in the southern hemisphere, such as New Zealand and Australia). The historical progression for industrial development has been that changes in technology, land

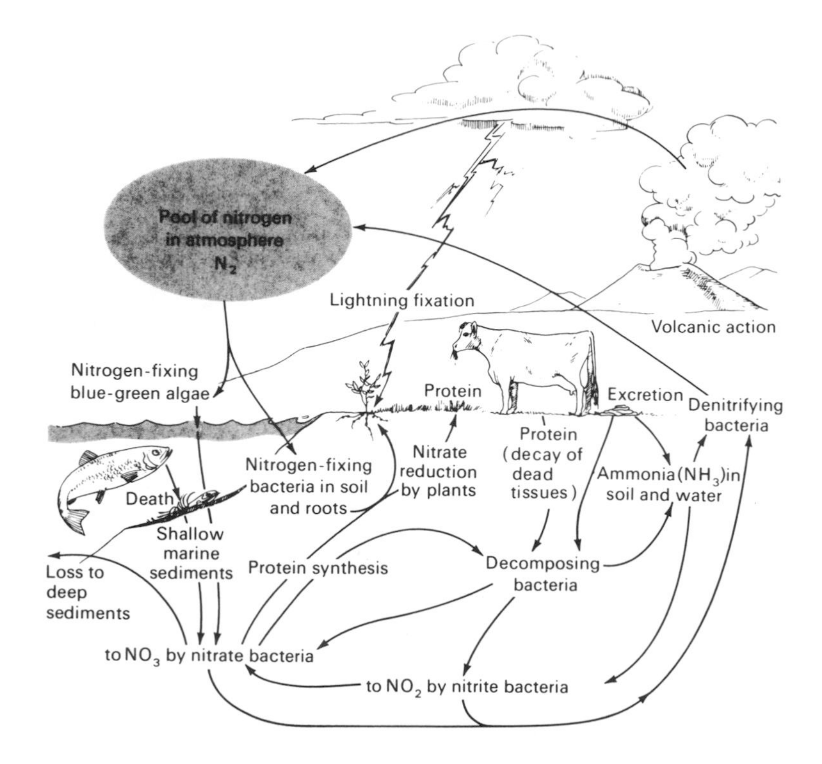

Figure 1.4 The nitrogen cycle. Source: Paul R. Ehrlich, Ann H. Ehrlich, and J. P. Holdren, *Human Ecology: Problems and Solutions* (San Francisco: W. H. Freeman, 1973), p. 154.

tenure, and cultivation practices occur in agriculture, leading to an agricultural revolution, which generates the demand for local, small-scale industry, and a surplus of capital, which becomes the base upon which urban industrialization is built. Thus, by providing a favorable climate for agricultural production, the temperate zones promote industrial development.

After World War II, attempts were made to help the poor countries of the world "develop"; however, the model used, one of rapid economic growth through industrialization, either ignored the need for prior rural and agricultural development or hoped simply to bypass it. The fact that countries in arid and tropical climatic zones face a range of climate-based obstacles to rapid agricultural expansion and that the experience of temperate zone agriculture is of limited value in these countries has meant that there has been

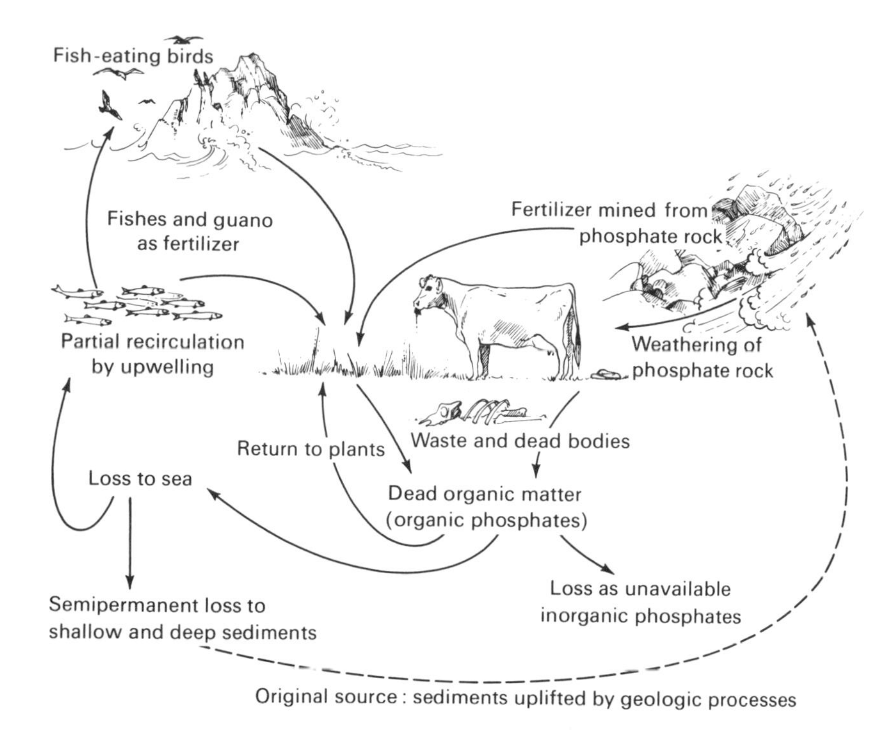

Figure 1.5 The phosphorous cycle. Source: Paul R. Ehrlich, Ann H. Ehrlich, and J. P. Holdren, *Human Ecology: Problems and Solutions* (San Francisco: W. H. Freeman, 1973), p. 156.

widespread agricultural development only when there has been significant social reform (especially land reform, as in the cases of Japan, Taiwan, and China) that has been combined with locally adapted and improved agricultural practices. Those areas where energy-intensive, industrially supported agriculture (the so-called Green Revolution) has been promoted have had increases in production, but mainly in the already relatively rich, irrigated regions.

In any case, the links between an expanding agriculture and the growth of industrial society over the past several centuries have meant that both forests and grasslands have shrunk as agriculture has expanded. When one compounds these trends with today's exponentially increasing global population, it is clear that mankind has reached the point of seriously wasting or destroying the natural "capital" upon which societies depend for their long-term sustenance.

Another dimension to the explosion of industrial society is the way in which it has transformed both individual societies and a wide range of larger systems. As indicated above, industry grew out of an agricultural revolution. Since then, there has been a steadily widening and deepening impact of industry upon society and the environment. The first stages of the industrial revolution involved the creation of individual industries, often large, sometimes competing, but more often merging into large monopolies whose power and influence then grew to involve entire sectors of industry. Today, the full-scale involvement of industry in the state and vice versa is recognized in the concept of a *national* economy, and in a wide range of practices whereby countries seek to assist their industries through tariffs, quotas, export subsidies and insurance, and financial assistance in times of difficulty.

In addition, an increasingly global industrial framework has been established. Trade patterns and networks, huge shipping systems, satellite communications networks, international banking consortia, exports and imports of all manner of industrial and agricultural goods, and the rapid growth of multinational (or transnational) corporations have all been parts of this expanding framework. There are a number of aspects of this global trend that have raised passionate debate. There are questions about the degree of centralized corporate control or exploitation that may result from expanded large-scale production of consumer goods. There are questions about whether the high energy and environmental costs of current industrial systems can be maintained very long into the future. There are questions about whether the multinationals and the various foreign aid programs coming from the industrial countries really encourage development or whether they simply lock the poor countries into a dependent relationship where they supply various raw materials while receiving some industrial goods and capacity in return.

To get some idea of the great changes that have occurred in the past century alone as a consequence of this global industrial framework, let us take two imaginary journeys: one a century ago in a balloon, the other today in an airplane. If one had floated in a leisurely way across parts of the United States in a hot-air balloon in the 1880s, one would have been able to see numerous examples of the impact of man and industry upon the landscape. One would see towns and cities. One would see some factories, steel mills, and mines. One would see farms and fields. All of these would be seen to be linked by an infrastructure of dirt roads, shipping canals, various ports on the coast filled with sailing ships and steamships, and the tracks of various railroads cutting across the landscape and linking the farms, towns, mines, and cities.

Today, in flying across the same parts of the United States, one will see not only all of the above, but also a huge and complex network of asphalt and concrete roads. Automobile factories, autos, trucks, parking lots, garages, oil refineries, and oil storage tanks are all linked into this road network. On the coast, there are giant supertankers bringing oil from the Middle East or Africa to huge ports in the Gulf where other ships are loading grain for export. There are immense hydroelectric dams, coal-fired generating stations, and nuclear power plants with their cooling towers, all plugged into regional power transmission lines and grid systems. One sees airports, shopping centers, suburban housing developments all covering over and replacing what had been prime farmland a century before. Mining operations include not only underground mines, but broad open-pit mines and long deep cuts into mountains and ranchland where gigantic shovels and trucks steadily strip mine thousands of acres for coal. In short, the landscape has been overlaid and transformed by an increasingly large and complex infrastructure that is energy and capital intensive. This infrastructure, of course, has a direct impact upon the environment when built and often continues to create environmental stress through pollution, erosion, or resource depletion during its operation.

There are several characteristics of this infrastructure and its accompanying institutions that are worth noting. First, they have been structured to maximize production. This means that all other considerations have been given a secondary place, including the social and environmental costs of operating them. Second, they are large in scale, and centralized in their overall control and operation. Third, the highly specialized and complicated structure of these large systems means that they can change their basic operations only slowly, i.e., they are rigid. In fact, it is this rigidity that increases their productivity. As the example of the U.S. auto industry shows, it took four to five years to shift production from an emphasis on large "gas-guzzling" autos to smaller, more fuel-efficient ones.

These characteristics are also reflected in the two larger trends, or explosions, that have been described. Both the population explosion and the explosion of industrial society have a great deal of momentum. That is to say, their structure and size make it a slow and difficult process to change their basic directions. The demographic profile of the populations of the developing countries—where often over half the population is under fifteen years of age—means that even if one could somehow magically reduce the birth rate to what is usually considered to be the "replacement" level of 2.2 children per couple, it would still take a number of decades for the overall size of the population to level off. Similarly, the momentum of industrial society is such that it will take many decades for it to be transformed. This momentum compounds our current environmental problems because most of them relate directly to either the increasing numbers of people in the world or to the increasing consumption of resources and the generation of pollution that flow from current patterns and practices of industrial society.

Having reviewed the different cultural views of nature, environment, and society as well as the broad historical trends that underlie the increasing threats to our biosphere, we need to turn to the difficult matter of just how we are to analyze, understand, and evaluate those threats. This is no easy task, as can be seen in a number of important debates, such as those on the risks of nuclear energy, the risks of climate changes resulting from carbon dioxide buildups, the long-term risks of pesticides such as DDT, or the dangers of toxic chemicals.

Problems of Evaluation and Forecasting

Environmentalists are sometimes accused of conjuring up doomsday or disaster if certain actions are taken or certain practices are continued. To the degree that this tendency exists, it has several sources. In the early days of the environmental movement, most people were unaware of the larger dimensions of the environment and had little understanding of basic ecological processes and the kinds of risks associated with their disruption. In this sense, crying "wolf" was a way of trying to get people's attention, to get them to consider the issues seriously. Of course, as the fable illustrates, there are serious dangers in crying wolf when a wolf is not actually at the door. On the other hand, many of the disasters that have been suggested are real possibilities, although no one is quite certain of the actual probabilities of their happening. These uncertainties flow from three basic problems that will be discussed below: the lack of basic data (or gaps in such data) regarding the critical characteristics of important ecological processes; the delays that tend to occur before the full impact of certain environmental actions is felt; and the difficulties involved in trying to get a good overview by using intellectual tools that are designed to look at narrow disciplinary questions—something akin to trying to view the landscape through a microscope.

Inadequacy of Data

The problems of the availability of basic data are particularly apparent when one seeks to understand global environmental problems. Whether one is dealing with questions of climate change, ocean pollution, deforestation, soil erosion, or whatever, there are two key types of gaps. First, complete or adequate data on current patterns are rarely available. Second, even in those few cases where there are some current global data, there is rarely any good historical evidence from which one can determine whether the current situation is typical or atypical. There may not even be any discernible patterns or trends in the historical record, which may appear simply as random behavior.

Let us take a few examples. No one has more than a general idea of what is occurring globally in terms of soil erosion and even less of an idea as to what that erosion means either for the global ecosystem or for specific regional ecosystems. It is not merely that many countries lack either the money or the capacity to gather statistics on soil erosion. Even if there were good global statistics on rates of soil erosion, they would say little about either the causes of that erosion or its consequences to various specific environments. Causes of soil erosion can range from unusually heavy or unseasonal rains, to the removal of ground cover through deforestation or agriculture, to the overgrazing of grasslands, especially by sheep and goats, to various kinds of construction activities such as road building, to diverse types of extractive industries, such as strip mining. The effects of soil erosion can also be different from region to region and from ecosystem to ecosystem. In the 1930s, important parts of the United States suffered the long-term consequences of careless farm cultivation practices that turned the Great Prairie into a "Dust Bowl." Other effects of soil erosion can include the development of deep gullies, the salinization or acidification of streams and ponds, and the rapid silting up of dams. This latter problem has been especially serious in several developing countries, where dams built on cost-benefit calculations of an eighty-year productive life span for the dam now face huge losses because the dam will be completely silted up in thirty to forty years.

The problems may be even more complicated in the case of climate. Even though what is now the World Meteorological Organization (WMO) was founded in the nineteenth century, both current and historical weather and climatic data are spotty. Historically, it was primarily the Europeans and the Americans who gathered systematic data for their countries. Very little was collected in Africa, Asia, and Latin America, and what was collected tended to be for the capital or port cities or for agricultural areas. Today, there are much better data being gathered in the many countries of the world; however, that still leaves the largest portion of the earth's surface—the oceans—largely uncharted in terms of systematic weather or climatic data. The development of satellites has been a tremendous aid in trying to follow atmospheric and weather patterns as they cross both the oceans and continents, but to be able to develop better understandings of these patterns, much more is needed in the way of "on ground" or "on water" reporting networks. In addition, atmospheric patterns and flows are clearly influenced by the various ocean currents. While the WMO has encouraged the development of better reporting networks and has built a large computer system to bring all these data together, such work is expensive and requires both the financial and organizational support of the many countries of the world, something not always easy to obtain.

Finally, it should be repeated that at the global level, these various systems—the land, water, forest, grassland, and atmospheric systems—all interact in the hydrologic cycle. Because of this, major changes in any one of the systems will affect the larger cycle and will have definite effects upon climate, which in turn will influence the functioning and relative balance of the component systems. Thus, one is in the uncomfortable position of knowing that there will be effects, but not knowing exactly what they will be or how serious their impact will be upon given areas or countries or peoples.

Delayed Impacts

The second major difficulty in analyzing environmental problems relates to the long time it takes for many actions or processes to produce their full results. There are both scientific and policy reasons for this difficulty. In scientific terms, it may take a number of years and a large number of expensive tests or experiments to analyze the effects of something like a new chemical compound upon the environment or upon humans. Even then, the results can usually only be expressed in terms of probabilities, in part because other factors may speed up or inhibit the effects of such a compound. Since this is the case, it then becomes difficult for policy makers or for courts to decide how to assess the environmental impact of the substance and take appropriate action. Cigarette smoking serves as a current example. The medical evidence is clear that smoking has a number of negative effects on human health. Also, the statistical evidence is clear that the probabilities of heavy smoking leading to lung or throat cancer are great. However, because not everyone who smokes heavily does get these cancers, it is clear that there are other factors at play—general health, the presence or absence of other pollutants, the degree of stress, and perhaps certain genetic characteristics. Because of this, and because the negative effects become serious only after twenty or thirty years, many young people ignore the warnings and think that they can "beat" the probabilities or that even if they do get cancer that some miracle cure will have been found by then. Although there have been some modest efforts at regulation, such as controls on the advertising of cigarettes, there have been few damage suits. This is because the warning label on cigarettes, which the industry fought against so strongly, has been interpreted ironically as relieving them of any legal liability for health damages resulting from the use of their product. After all, the individual still chose to smoke even after being warned.

In situations where individual choice is less of a factor, there are more opportunities to seek legal or political redress. Even so, the scientific uncertainties often make this difficult. A particularly unpleasant set of examples relates to the use of dangerous, toxic, or cancer-causing materials in the workplace. For example, asbestos has been widely used in industry to make clutch and brake linings, asphalt tiles, wallboard, furnace linings, etc. Starting in the 1960s, studies began to show that workers exposed to even small amounts of asbestos dust over the years had much higher cancer rates than normal. As the scientific evidence has become more and more clear cut, the battles to protect workers' safety have increased. However, safety costs money and not all corporations have been eager to institute expensive safety measures. As these have been increasingly mandated because of congressional and regulatory pressure, we have seen instances of what is becoming a fairly common phenomena in the environmental health field: the relocation of such factories to developing countries where hazardous practices can be carried on with few health, safety, or environmental restrictions and where, in addition, labor is cheaper. American workers are then caught between the desire for a healthier and safer workplace and the threat of the loss of jobs if their company thinks that safety procedures would be too expensive.

One of the most complicated legal proceedings in recent times has been the so-called Reserve Mining Case. For a number of decades Reserve Mining Company (equally and totally owned by Republic Steel and Armco Steel Corporations) has been the largest single source of pollution in the U.S., dumping some 75,000 tons of taconite tailings (wastes) into Lake Superior each day. Because it got the first and only permit given to dump such wastes into the lake, it has enjoyed a significant economic advantage over the other fourteen taconite mining firms, which have had to dispose of their wastes through more expensive on-land procedures. Besides the pollution and the damage that these wastes cause to fish and wildlife, there emerged in the 1970s the question of whether the asbestoslike fibers in the wastes might constitute a health hazard to those in Duluth, Minnesota, and elsewhere in the region who were drinking water containing these fibers. The dilemma for the courts in terms of the health question (there were many other important legal questions as well) was whether the health risk of these fibers (possibly causing cancer to a number of people in twenty to thirty years) was great enough to require Reserve Mining to stop its dumping of wastes into Lake Superior. The lack of clear-cut scientific evidence eventually led the courts to require a combination of filtration of city water supplies and eventual movement of the dumping onto a land site.

As one moves to more global types of processes and risks, the questions of how to assess and deal with them increase. Not only are there the scientific uncertainties, but there are different impacts on different groups, often in separate countries. There are a number of examples, many relating to airborne pollutants. One early controversy raged around the atmospheric testing of nuclear weapons, both in the United States (New Mexico and Nevada) and in the Pacific. Initially, the physicists working for the Atomic Energy Commission assured the public that there were no risks from such fallout because there was no way for people to become exposed to any significant amount of radiation. This turned out to be incorrect. Ecologists found that radiation was being concentrated in the food chain with dangerous levels appearing in both fish and milk. A similar controversy continues today over the risks of low levels of radiation. Official spokesmen for the Nuclear Regulatory Commission and the nuclear industry claim that small releases do not endanger public health. Some scientists, however, claim that there is no safe level of radiation and that every exposure over natural background levels will have some harmful effect.

Similar uncertainties abound in the controversies surrounding the long-term effects of fluorocarbons on the ozone layer. These fluorocarbons, sometimes known by the trade name, "freon," are used as propellants in aerosol sprays and in air-conditioning and refrigeration systems. It is roughly an $8 billion industry in the United States. The fear of scientists has been that these fluorocarbons, which rise very slowly into the upper atmosphere, will gradually reduce the ozone layer. This layer shields the earth's surface from various types of radiation, especially ultraviolet radiation. Increased levels of radiation would lead to increased levels of skin cancer among humans and would have various effects upon plant growth, animals, and fish. Another possible impact would be changes in regional climate flows and temperatures. The difficulty here (again) is that it will take some ten to fifteen years to find out if the predicted possible effects are accurate and if they are, it will then be too late to do anything about them or the large reservoir of fluorocarbons still rising up to the ozone layer. There has been legislation in many industrial countries to reduce or ban the use of fluorocarbons in aerosol cans. Although it has been relatively easy to reduce this source of pollution in such a nonessential arca as aerosols, the problem of dealing with the use of fluorocarbons in refrigeration systems is much more difficult. Scientific studies, which at first were divided about the degree of risk, now are largely concluding that the risks are greater than originally thought. As a result, the U.S. Environmental Protection Agency announced that it would seek to hold U.S. production levels constant and would seek similar agreements from other industrial countries.

Specialized and Partial Approaches

The third major difficulty involved in analyzing environmental problems results from the fragmented and specialized nature of most of our institutions: governments with their departments, bureaus, and offices; corporations with their divisions, departments and branches; unions with their national, regional, state, and local offices; and universities with their colleges, institutes, centers, and departments. All these analytical and jurisdictional specializations are quite in contrast to the integrated nature of the various ecosystems that have been described. Unfortunately, it is not simply a matter of fitting together whatever jigsaw puzzle pieces these specialized groups produce. They are normally working on different "pictures" or problems. Even if they are working on the same problem, their differences in perspective and their institutional vested interests make true integration of the pieces difficult. The situation parallels the story of three blind men trying to describe an elephant. One has hold of the tail and describes the elephant as a snake; the next has hold of the leg and describes it as a tree; and the third has hold of the ear and describes it as a sail. The point is that each is sensing only one small part of the larger reality and therefore simply describes the whole in terms of the small part to which he has access. Our academic disciplines and scientific fields suffer from similar problems. We have the great divisions among the natural sciences, the social sciences, and the humanities. Within them, we have numerous disciplines, each focusing on one small part of reality (although each discipline normally sees itself to be the "key" or "master" discipline).

Another aspect of this specialization is that it has been closely linked to the growth of modern industrial society and its needs. Thus, the university has gradually been transformed; it is no longer a separate institution in society. Today, universities include large professional schools of engineering, dentistry, law, medicine, etc., which have as their main purpose the application of knowledge rather than the generation of new knowledge. Also, many of what used to be purely "ivory tower" fields, such as nuclear physics, are now strongly linked into both industrial programs and large government grants. The institutional growth of environmental and ecological programs in such a context has been difficult, even though there has been recognition of the need for interdisciplinary and comprehensive research, analysis, and educational programs.

It is apparent that our best estimates of environmental risks, threats, and limits, are far from complete or adequate. However, they do represent major efforts and one must take them seriously. Another major point relates to the dominance of economic modes of thought and analysis in most present day policy making. The discipline of economics, although it has an important tradition, is still only one specialized discipline and it captures only one small part of reality. It has itself chosen to ignore (or consider as "external") such things as the environmental and social costs (and benefits) of various industrial processes. Also, it has tended to ignore basic value questions regarding the importance of the environment and society's relation to it.

What are needed are new ways of evaluating the full costs and benefits of various proposed programs and policies. In some ways technology assessment and the use of environmental impact statements have helped in this regard, but they still suffer from the larger problems outlined above: lack of adequate data; problems of understanding and predicting long-term effects; and specialized thinking. Technology assessment and environmental impact statements have been used with most success on smaller-scale projects and in the more industrial countries. However, when it comes to making assessments of global risks and limits, the above difficulties apply with full force. Even so, it is important to try to outline those limits and threats as best we know them, recognizing that they are far from complete. As with many international problems, this simply means that one must make the best judgments possible with the information available, keeping in mind the great virtues of caution and prudence in such circumstances.

Global Environmental Threats and Limits

In this section we shall try to summarize many of the trends and threats that confront the environment. The threats will be discussed roughly in terms of the four basic elements that the ancient Greeks saw constituting the physical world—air, water, earth, and fire. Threats to global atmospheric and hydrologic systems will be summarized first. We shall then address the loss of cropland, rangeland, and forests, followed by an analysis of the loss of the world's living resources and their habitat. The threat of nuclear war will then be reviewed. Finally, we will turn to the question of limits—both resource and social—which represents a major challenge to modern industrial society.

Atmospheric Systems and Climate Change

As suggested earlier, changes in any of the basic ecosystems making up the hydrologic cycle will tend to be reflected in either regional or global changes in climate. Let us now examine some of the general risks and threats associated with climate changes. First, it should be pointed out that there have been significant climate variations throughout history, with noticeable ones occurring over the past thousand years. Also, it should be noted that there appears to be no regular pattern or cycle associated with these changes. This phenomenon must be due, in part, to the many things that can influence climate. Volcanic dust is one such influence, but there is no clear pattern of volcanic activity, especially the very large eruptions that can affect the climate for several years. Sunspot activity appears to have some influence; there is some correlation in the United States over the past couple of centuries between droughts and the eleven and twenty-two year sunspot cycles. However, other influences, such as air pollution, may be more important and may either cancel out or reinforce the effects of sunspots.

At a global level, probable changes in the average global temperature are of great concern to scientists. If there were a significant decrease in the global average temperature, then the world, and especially those of us in the northern latitudes, would be confronted with a new ice age. This would have several major effects. It would drastically reduce agriculture in the temperate zones—those regions that have been the most productive during modern history. It would require dramatic changes in the housing and employment of millions of people in the industrial countries. And there would be no guarantee that the climate changes in the arid or tropical areas would benefit their agriculture significantly. Thin and fragile tropical soils would remain, even though a more moderate climate theoretically would permit different crops to be grown.

At the other extreme, any significant increase in average global temperatures would tend to melt the polar ice caps with a resulting rise in the level of the world's oceans and seas. The result would be an inundation of low-lying coastal areas and cities. A rise of fifty feet would threaten many of the world's major cities. While there might be some improvement in growing conditions for agriculture in the northern latitudes, conditions in the equatorial regions might deteriorate. Thus, as with many other global environmental problems, the impacts would differ significantly between various regions and countries. Some might benefit, while many others might suffer. One of the curious dilemmas of modern industrial society is shown here: once the normal range of temperature and climate limits within which industrial societies have developed are exceeded, they then become more vulnerable than many traditional or developing societies. This is because the latter are more closely linked to their environment and are more used to adapting to changes within it than trying to overcome them through technological means such as houses, heating systems, air conditioners, snow plows, large farm equipment, specialized hybrids, etc.

It is the increasing rate of carbon dioxide (CO_2) concentrations in the atmosphere that most concerns scientists in regard to a global increase in temperatures. It is known that this increase creates a "greenhouse effect" where the solar radiation that enters the atmosphere is less able to reradiate into space. The more CO_2 builds up, the more temperatures will increase—assuming that other things are equal (which they are not). There has been a historic CO_2 buildup over the past century, especially as the amount of fossil fuel burned has increased (and it is expected to continue to increase still further). These fossil fuels represent one of the two great pools of carbon reserves on earth. They are the remnant of millions of years of plant and tree growth, which mankind is using up in just a few centuries. The forests and soils of the world, the other great reserves of carbon, are also suffering from serious depletion. The relationship of deforestation to soil erosion has already been discussed. In terms of climate, large scale deforestation threatens to increase the amounts of CO_2 in the atmosphere as well as to change the albedo of the earth—that is, the relative balance of those areas that reflect light and heat (especially the deserts) and those, like the forests, that absorb light and heat. A change in albedo could well have regional climatic effects and perhaps some global impact. The CO_2 question is linked not only to forests, but to the oceans. The oceans serve both as a sink for CO_2 and as a major source of the world's oxygen. Thus, any marked changes in the capacity of the oceans to absorb CO_2 or to produce oxygen will potentially affect both the hydrologic cycle and climate.

Hydrologic Systems and Water Pollution

Another broad but related global problem area is that of ocean and lake pollution. The source is simple: people, especially in Western cultures, have historically tended to see rivers, lakes, and oceans as a dumping ground. All manner of wastes are dumped into them. In recent decades there have been modest efforts to process or limit what can be dumped into the world's waters. Most of the effort has been focused on inland waters where one country has jurisdiction and where it suffers the consequences of its own water pollution.

One of the main problems with global water pollution is that it is difficult to study the effects of ocean pollution. There is, generally speaking, much less overall scientific knowledge than for land ecosystems. We do know some of the more direct and drastic effects. Heavy metals, such as mercury, have been shown to concentrate in fish and shellfish causing serious health disorders or death for those who eat them regularly (as happened to over one hundred people living next to Minimata Bay, Japan). And in the Great Lakes, there is virtually no game fish that is not contaminated with one of several pollutants—either heavy metals, pesticides, or PCBS.

These various pollutants have an impact upon phytoplankton, the primary producers of protein in water systems. DDT especially has been shown to have been dispersed over virtually the whole globe—with measurable amounts found in Antarctic fish and penguins. DDT, though banned in the United States, is still used in many other countries as an easy and inexpensive (but very shortsighted) way to try to deal with malaria. Unfortunately, more and more mosquitoes and other pests are becoming resistant to DDT and other similar pesticides. More important, DDT enters the food chain and has had a damaging impact upon those birds that feed upon fish. Another major source of ocean pollution is oil—whether from oil tankers or blown-out oil wells. Damage is often very high, whether to marine life or to beach-based tourist industries. While international organizations have been working on questions of liability for such spills for many years and there are at least some generally accepted principles, specific cases are often difficult to settle. Also, liability for damages relates only to the most immediate and obvious forms of damage; scientists still do not know all of the long-term consequences, particularly on a global scale.

Another major source of water and ocean pollution is land-based industry. A great deal of what is usually classified as air pollution eventually "falls out" into some body of water. Thus, it appears that most of the lead appearing in the Great Lakes has come from automobile exhaust and factory emissions. A recent development of concern is what is termed acid rain. High levels of sulphur dioxide (SO_2) in smoke increase the acidity of precipitation, but often several hundreds of miles from the sources of pollution. Thus, New York and eastern Canada suffer from the acid rain generated by the steel factories in Indiana and Ohio. The Scandinavian countries suffer from the acid rain produced by the smoke of British steel mills as well as those in the Ruhr area. The effects have been to reduce the growth of forests, especially pine forests, and to reduce or kill the fish in a number of lakes. Also, many buildings and monuments have been seriously defaced.

The Loss of Cropland, Rangeland, and Forests

The capacity of the earth to provide the food, fiber, and wood that is basic to all human societies is being seriously undermined by various types of losses. Soil erosion losses are hard to estimate on a global scale, but are seen as threatening by all observers. The situation in the United States is serious and one must ask what that must mean for other countries that are less well endowed with soils, are poorer, and have no Soil Conservation Service (SCS). A recent survey of farms in the main grain growing areas of the United States found that 84 percent had annual soil erosion losses above the natural rebuilding rate.[3] A 1975 report by the Soil Conservation Service found that soil losses on U.S. cropland amounted to almost three billion tons, an average of about nine tons per acre. The report concluded that to be able to sustain production at current levels, soil losses would have to be cut in half, something that would require major SCS policy changes and a 5 to 8 percent increase in production costs.[4] Unfortunately, the short-term economic interests of farmers and urban consumers make it politically difficult to take the actions needed to preserve the basic long-term health of our soil systems.

Another source of significant loss of productive capacity is the growth of desert areas—what is called desertification. This occurs much more in selected areas—where land is abused—than it does along existing desert borders. Most desertification occurs on rangeland that is overgrazed. Overgrazing results from increasing livestock populations (stemming from both an increased demand for meat and the cultural preference of nomadic tribes for more livestock as a status symbol), from the conversion of rangeland into cropland (leaving less rangeland per animal or pushing grazing onto more fragile lands), and from natural causes such as extended drought periods. All of these factors were found in the 1968–73 drought in the African Sahel, the drastic consequences of which led to a United Nations Conference on Desertification in 1977. It has been estimated that if all the high-risk areas identified at the conference became deserts by the year 2000, deserts would occupy more than three times the 7,992,000 square kilometers they occupied in 1977.[5]

Two other factors that significantly affect the productive capacity of crop- and rangelands are changes in soil quality and the taking of these lands for other uses (what is usually termed conversion). Changes in soil quality normally result from some sort of overuse. The organic matter or humus in soil can be lost by continual use of single crops (monocultures), particularly corn and soybeans, through the burning of crop residues, or through their use for fuel. These losses have been masked by the use of chemical fertilizers and will become more apparent as fertilizer use declines due to increasing cost. Overuse of a different kind has resulted in serious problems of waterlogging, salinization, and alkalinization. These problems are most serious in arid regions where large irrigation projects bring in much more water than is typical for that climatic zone. If current rates of cropland lost to production due to these water-related factors continue through the year 2000, about 2.75 million hectares (approximately 1.4 percent of the world's total irrigated land) will be lost to production—an amount of land that could feed between 9 and 15 million people.[6]

Conversion of cropland to other uses—largely urban sprawl—constitutes an even greater potential loss, particularly because the land converted is very often prime agricultural land—that which is level, has deep soils, and is highly productive. In the United States, 23 million acres of agricultural land (which includes rangeland and forests) was converted between 1967 and 1975.[7] Estimates on a global scale are that if current trends continue, urban development and conversion will claim 25 million hectares of cropland by the year 2000, enough to feed some 84 million people.[8]

Global deforestation carries with it three different kinds of threats. First, it may cause climate changes in ways that have already been discussed. Next, the world's tropical forests are great reservoirs of genetic diversity, which we can ill afford to lose for reasons explored below. Finally, deforestation greatly increases both soil erosion and flooding. Current rates of deforestation are staggering. Estimates suggest a loss of some 50 million acres per year of "closed forest" (where there is a tree canopy covering the ground). This means an area roughly the size of the state of Nebraska is being lost each year. By the year 2000, the projections are that the world's closed forests will be reduced from one-fifth of the earth's land surface to one-sixth. The estimated losses for open forest are some 12.5 million acres per year.[9] Besides the destruction of habitat and the lost genetic diversity that such deforestation represents, it also increases the likelihood of shorter-term economic losses, such as the 1978 flood of the Ganges River in India. The severity of this flood, which was the worst in one hundred years, was blamed in good part on the deforestation of the watersheds in the Himalayas. The flood cost some 2,000 lives, killed some 40,000 cattle, and inundated two provinces at a cost of at least $750 million in crop losses.[10] The dynamic interactions of cropland, rangeland, and forest loss are compounded by population growth. We may have reached a stage where resource use per person is perhaps more important than pure numbers in terms of environmental impact.

Species and Habitat Loss

There are a number of threats that involve the loss of genetic materials and genetic diversity. First, it should be noted that the number of species that have become extinct has increased at a geometric rate, as have the underlying causes for such extinction: the explosions of industrial society and population. Animals becoming extinct have received the most attention, especially with the publicity given such large and exotic endangered species as Siberian tigers. However, the loss of plant species may be more significant because the extinction of a plant can lead to the loss of another ten to thirty dependent species such as insects, animals, or other plants.[11]

The lack of basic data is especially visible in matters of genetic diversity. It is estimated that only somewhere between 15 and 20 percent of all species have been recorded in the scientific literature; many species have yet to be discovered, and ironically may be lost before being recorded or studied. And it is in the tropics where there is the least knowledge, the greatest number of species, and the greatest deforestation. "Whereas a European forest typically contains 5 to 10 species of trees per hectare (2.5 acres), a tropical moist forest usually has 100 to 150 different tree species—in addition to a rich diversity of other plants and animals."[12] As botanist Peter Raven has observed: "Billions of dollars have been spent on the exploration of the moon, and we now know more about the moon than we do about the rainforests of, say, western Colombia. The moon will be there far longer than these forests."[13]

But why, one might ask, should we be so worried about the loss of various plant, animal, insect, and fish species? There are a number of reasons for concern. Let us start with the least important, the short-term pragmatic reasons, those primarily of economic benefit. The loss of various fish species due to overfishing leads to the loss of jobs on the part of many small fishermen and makes the task of catching and processing more difficult and costly. Also, of course, species loss does something that is more than economic; it reduces the availability of easily obtainable and relatively cheap protein for many poor countries and poor people. More serious are the losses of various traditional varieties of food plants. These species, which have been gradually improved over the millenia by careful observation and selection by peasants, are being displaced by the increasing use of high-yielding hybrids. These hybrids, and the capacity of plant breeders to stay ahead of continually evolving varieties of plant pests and diseases, depend upon the availability of a wide range of genetic material with which to crossbreed. While seed banks are used to reduce some of the risks of the loss of traditional and wild plant-breeding stocks, they are a second-best approach to in-field preservation. The economic losses that can result from major pest or plant disease outbreaks are obvious, as was shown in the 1970 outbreak of corn blight in the United States, when some 15 percent of the total crop was lost. The same sort of considerations apply in forestry, where tree breeding is becoming increasingly important.

Another economic area where plant availability is important is in pharmaceuticals. Few people realize that roughly 25 percent of our prescription drugs contain agents derived from plants, while another 15 percent contain agents of microbal or animal origin.[14] Also, there is an increasing amount of research into plant biochemistry, particularly in the search for treatments for cancer. Finally, there is greater interest in various folk and traditional herbal remedies, many of which have been shown to be based on plants with valuable biochemical properties. To reduce the available pool of genetic materials from which to draw crops, tree species, industrial chemicals and dyes, and pharmaccuticals is shortsighted, even in conventional economic terms.

In a deeper sense, however, the loss of species can be seen as a sort of early warning system, alerting mankind to future dangers, just as miners used canaries to detect dangerous gases in the mines (the canary would be overcome long before the miners and thus provide warning). At a global level, one must be aware not only of the individual species that are being lost, but the underlying causes —which all take the form of either loss of habitat or ecosystem simplification. Loss of habitat is most visible in the increasing disappearance of tropical forests. The main causes of this are commercial timbering, the clearing of land for agricultural purposes (where such cleared land in the tropics can often be used only for a few years before it hardens beyond the point of use), and firewood gathering. Each of these is related in large part to the poverty of the people and countries where these forests are located. Thus, any long-term approach to saving these tropical forests must be based upon genuine rural development, recycling systems, and what are termed appropriate technologies.

Ultimately, there are the questions of what sort of world mankind wants to live in, and whether he is willing to do what is necessary for his own survival as a species. Although man may have certain technological means to delay or temporarily cushion the effects of certain natural laws, he is not exempt from the general principle that as ecosystems are disrupted and simplified, it is the species at the top of the food chain that become the most threatened and vulnerable. Western man is beginning to realize that the finiteness of fossil fuels will mean that modern industrial society, which has been built upon those fuels, will have to change. Western man has still not realized the wisdom represented both in many non-Western cultures and in the most advanced scientific thinking on environment and ecology: that human life itself depends upon the healthy functioning of the earth's biological systems. In terms of the discussion on cultural differences, Western man is going to have to learn that he is *part* of nature, not superior to it, that he must respect his cohabitants and the great physical systems that we all share. Otherwise, we seriously limit our future evolutionary choices and options and risk the very existence of the human enterprise.

The Effects of War

A final broad category of threats to our global ecosphere is perhaps better known than the others, that is, threats arising from military activities, especially the risk of nuclear war. Usually, war is thought of as a political phenomenon, not as a global environmental threat. This is a prime example of the dangers of overspecialized thinking. War-related activities not only threaten the global ecosphere, but involve many social, political, economic, and local environmental threats even if nuclear war is averted. Also, it has become increasingly clear that one cannot really separate "military" and "peaceful" nuclear power. The expansion and diffusion of so-called peaceful uses not only carries with it health and environmental risks, but also the risk that the by-products and waste materials from these peaceful reactors will be diverted to produce nuclear weapons, which might be used in local conflicts or for blackmail. Nuclear reactors and weapons are not the only war-related threats; there are also those posed by the development of chemical and biological weapons.

The fact that the mushroom-shaped cloud of a nuclear explosion is easily captured on film or television has helped to make the average person aware of the magnitude of possible destruction that nuclear weapons can wreak. Even so, it is hard to imagine the full consequences of a nuclear attack. In an effort to do so, the U.S. Congress requested that its Office of Technology Assessment (OTA) prepare a report on the subject. In its report, OTA presented a number of scenarios. First, the effects of different weapons used against two cities are illustrated using the examples of Detroit and Leningrad. The report gives the following estimates.[15]

1 Megaton surface burst:	220,000 fatalities
	420,000 injuries
1 Megaton air burst:	470,000 fatalities
	630,000 injuries
25 Megaton air burst:	1,840,000 fatalities
	1,360,000 injuries

Although there are presently no 25 megaton weapons in nuclear arsenals, such a weapon dropped on the United States would leave virtually no habitable housing in the whole Detroit area and would destroy virtually all heavy industry. This is just one large nuclear bomb in one city. If one assumes a major attack using most of each side's nuclear arsenal, the immediate deaths are estimated to range from 20 to 160 million people, depending upon the time of day of the attack, the amount of advance warning, the wind direction, etc. Thc long-tcrm radiation effects during the next forty years would lead to the death of tens of millions more, plus many nonlethal cancers and other genetic disorders.

Of course, a large-scale nuclear war would affect much more than human populations. Plants, animals, forests, fields, and fish would all suffer the immediate blast effects as well as the longer-term effects of radioactive fallout and low-level radiation. There is little scientific data upon which to try to calculate the effects. It is definitely possible that there would be irreversible effects to the environment and various ecosystems. In regard to agriculture alone, there would be a double threat. First, the huge and complicated infrastructure and institutions that have made possible the high levels of production that we enjoy would be destroyed in a large-scale nuclear attack. On those lands not too severely contaminated by fallout, only rather simple growing techniques would be possible. Second, mutations or other basic changes in food crops or associated ecosystems might also reduce production or require major adaptations in cultivation or other farming techniques. Moreover, much of the world's agricultural expertise would be lost.

Another area of concern relating to large-scale nuclear wars has been the effect on the ozone layer. The amount of depletion would depend upon the number of large warheads used and the altitude at which they were exploded. Development of multiple warhead missiles with smaller warheads suggests that the risks are probably significantly lower than the 30 to 70 percent ozone reductions estimated in a 1975 National Academy of Sciences report.[16] Even so, and as is the case with most of the other global effects, the extent and seriousness of possible long-term consequences are enough to distinguish nuclear war from previously known wars. Past wars may have had high human casualties and serious local environmental damage; a large nuclear war would seriously disrupt the future prospects of the human species and would seriously threaten or destroy many other species and ecosystems.

Other environmental risks associated with war-related activities include the amounts of nonrenewable resources and fossil fuels that they use (even in times of peace), and more important, the risks associated with exotic chemical and biological weapons. Chemical weapons were developed as agents of war a long time ago. Poisonous gases were used in World War I, but not in World War II (partially because it was a more mobile war where the use of gas was even more difficult than usual). There are now available a range of gases, some of which, such as tear gas, cause only temporary incapacity while others, the so-called nerve gases, can cause quick death after exposure to only minute amounts. These nerve gases have not been used in war, but careless testing by the U.S. Army led to the death of over 6,000 sheep in Utah when test gases were released at too high an altitude. Also a range of deadly biological weapons has been developed, but there has been little inclination to use these in war because of their unpredictable and indiscriminate effects. In 1969, President Nixon unilaterally renounced the use of biological weapons. At the same time he submitted to the Senate the so-called Geneva Protocol prohibiting the first use of chemical weapons; however, he attached an "understanding" which was said to permit the continued use of "riot control" gases and herbicides in Vietnam. The widespread use of herbicides and other environmental weapons, such as the clear cutting of forests, represents one of the more unfortunate chapters of our whole Vietnam involvement because the effects upon both people and the environment will continue for several more decades, long after our withdrawal.

Resource Limits

So far we have examined various global threats to the environment. Let us now turn to the question of limits. In one sense, we have already discussed a number of limits along with the global threats, for each threat in one way or another involves certain environmental limits, which if exceeded lead to serious losses or suffering. There are two remaining and more commonly styled limits: resource limits and social limits. Questions and debate about resource limits have focused primarily on energy and the use of fossil fuels, although a number of other critical and strategic materials have been mentioned. There are two reasons for this focus. First, it is the nonrenewable resources that are at the base of modern industrial society. Clearly, there are immense practical and economic consequences if we run seriously short of any of the basic materials required by industrial society. The second reason relates to those Western cultural biases that tend to place nature (and thus all renewable types of resources) in a secondary status.

One of the main areas of controversy regarding nonrenewable resources concerns how one best estimates their present and future availability, on the one hand, and future need (or demand for them), on the other. Both types of estimates are fraught with uncertainties and one must make a number of assumptions that may or may not be valid. These include not only technical assumptions regarding the nature, concentration, and extent of different minerals, but also economic assumptions about the relative prices of these and other resources, and about the pace and cost of technological extraction, processing, use, and disposal, including innovations in each function. On the demand side, one must make assumptions about whether past patterns of use (at low prices) will continue as prices increase; whether there will be shifts in consumer preferences—for example, from large cars to smaller, more fuel-efficient cars; whether there will be government intervention either in form of restrictions or of subsidies; and so on. Also, though most estimates are based on the assumption that modern industrial society will continue to operate largely as it does at present, it is important to examine the possibility of basic changes in values, behavior patterns, and institutions. For, as has been suggested above, it does appear to many that we are moving into a period of transition where many things can be expected to change significantly.

There are continuing efforts to provide estimates of the remaining reserves of various critical and strategic minerals and fuels.

Normally, estimates are done on a mineral-by-mineral or fuel-by-fuel basis (for example, tin and copper, or oil and coal). Often these estimates serve to counteract a common, though unwarranted assumption: that as the higher-grade and easily extractable ores are removed for each mineral, there still remains a large proportion of lower-grade ores that may be less accessible. This is the case for several minerals, but not for others. And for many, though reserves do exist, the lower-grade ores may be prohibitively expensive to extract. Phosphorous, for example, which is a critical component of fertilizer, exists in granite formations but cannot be efficiently extracted from them. Similarly, there is a significant cost difference between drilling for the high-grade pools of oil found underground, and extracting the low-quality oil found in shale or tar sands.

Minerals and resources, though generally used individually, are part of a complicated industrial production and consumption system, where there are many possibilities for substitution, technological shifts, and value changes. High dollar or energy costs in extraction and processing, for example, lead to strong incentives to develop recycling systems. The example of recycled aluminum cans illustrates this phenomenon. However, the larger point that has only occasionally been addressed is how these multiple systems operate globally. One study that attempted to do this, and generated much controversy in the process, was called *The Limits to Growth*, which was sponsored by the Club of Rome and involved a team of MIT computer scientists.[17] Their World Model 3 was designed to try to project how five major systems interacted. The model used various complex feedback loops to represent the interactions among: (1) population growth, (2) the depletion of nonrenewable resources, (3) levels of pollution, (4) levels of industrialization, and (5) capital accumulation. The model suggested that collapse of the world system cannot be avoided by the leveling off of any one or two factors—for example, population or pollution levels—but only by a leveling off of all five. The value of the *Limits to Growth* report has been to demonstrate the complex interactions of different systems operating at dissimilar rates; in short, it demonstrates the interconnectedness of our global problems. The report has two significant weaknesses. It does not give much insight into how we might try to move toward a more sustainable system (something it really does not attempt to do). Nor does it take account of the various ways in which its global averages will be distributed or distorted in the real world.

It is the distribution of these various elements and resources that underlies a great deal of the debate about the "New International Economic Order" sought by the poorer countries in the southern hemisphere. The geographic accidents of the location of the world's largest oil fields as compared to the points of highest industrial demand illustrates just one such distribution issue. Another relates to resource use, where, for example, the United States, with approximately 6 percent of the world's population, uses roughly 30 percent of its nonrenewable resources. Because it appears unlikely on a number of ecological, environmental, and energy grounds that even the current world population could live at the U.S. standard of living without destroying the globe's supporting ecosystems, the debate over the new International Economic Order becomes one of redistribution as well as of redefining what is meant by "progress" or "development."

Social Limits

"Social limits" is a concept referring to the inability of different societies and cultures to change their habits, institutions, values, and patterns of behavior to accommodate major constraints. For the past thirty years especially, the question of social change has been focused on the developing countries in the Third World. The popular Western conception is that these societies are very traditional and that within them social change toward more rational and scientific approaches is slow and difficult, whereas Western society is much more rational and flexible. The first part of the stereotype is probably fairly accurate. The second part has been shown to be increasingly questionable as Americans and Europeans continue patterns of behavior that are both uneconomical and unecological. Although there has been an evident adjustment to the constraint of the oil-dependent global energy system, that adjustment has been slow. If individuals are slow to change their energy-wasteful habits and behaviors, our institutions are even slower to adapt.

Such rigidity and lack of adaptiveness on the part of our large, centralized, and bureaucratized institutions have made a number of observers skeptical whether industrial society will be able to adapt successfully to the predictable crises related to energy, resources, and water, not to mention such possible threats as climate changes, major wars, or epidemics. Industrial institutions and infrastructures are rigid and unadaptable outside certain fairly narrow limits precisely because they have been designed to achieve maximum production. The production assembly line, which symbolizes this sort of organization, requires strict timing, discipline, specialization of labor, plus huge amounts of energy to operate. It is a highly productive system, but slow to adapt to change. It recently required close to five years simply to shift production from large cars to small cars. To shift to other types of engines, for example, to electric or steam, would take even longer. Of course, much more than technical considerations are involved. Large units develop a whole set of vested interests around them that make it difficult for government to take what may be appropriate action or regulation. A sad, but unfortunately typical example was when President Carter, in a televised speech about energy, said in effect that a set of policies encouraging conservation would be in the national interest, but would not be approved by the Congress because of the power and opposition of the large oil companies.

An increasing number of analysts are suggesting that modern industrial society needs to become less energy intensive, less specialized, and less centralized; and that only by becoming so will we reduce the threats to our global ecosystem as well as increase the flexibility of industrial society to a point at which it may be able to adapt to the full range of probable changes and fluctuations. The question of social limits is crucial in any such large-scale transformation toward some sort of postindustrial society. The efforts made and the experience gained by citizens, governments, various professional groups, and international organizations in fashioning new and creative institutions to deal with the world's environmental problems have a direct bearing upon these social limits.

Responses to Global Environmental Threats

The international arena has been characterized by a wide range of rapid and difficult changes since World War II. The infant United Nations struggled with the new challenges of the atomic age and the emerging cold war. To the tensions and struggles of the cold war were added the pressures and difficulties of decolonialization, which peaked in the 1960s. The struggles for political independence and sovereignty have now been largely replaced by calls for greater economic independence, development, and justice. These political and economic struggles have been reflected in the foreign policies of the major powers as well as in the evolution of the United Nations and its specialized agencies, such as the Food and Agriculture Organization, the World Health Organization, and others. The gradual growth of both an awareness of and concern for global environmental matters in such a climate of controversy was not at all expected. The image of the earth as a serene blue planet floating in space, an image which could be spread over large parts of the earth through television and other visual media, certainly has added a large measure of awareness. But how does one explain the rapid growth of the environmental movement, the willingness of various citizen groups around the world to tackle long-ignored problems? There are no fully adequate answers, especially when one seeks to understand the interactions between actors at the local, regional, national, and international levels. What the remaining chapters attempt to do is to give the student an awareness of the various actors, values, policies, and futures that can be observed at the international level. Clearly, many of these are reflections of national or subnational concerns and vested interests; however, a range of programs, policies, and principles has been developed at the international level which have and will continue to influence governments, corporations, citizen groups, as well as the lives of many rural peasants who are not effectively represented in any forum.

A critical turning point in the increasing recognition of responses to global environmental problems was the United Nations Conference on the Human Environment held in Stockholm in 1972. This conference, which is examined in detail in chapter 2, brought a whole new series of actors into the international arena. Traditional governmental actors and intergovernmental agencies were confronted with a broad new range of unfamiliar issues and a new mixture of activist groups seeking real action, not just diplomatic resolutions. The Stockholm Conference created a new international actor, the United Nations Environmental Program (UNEP), which has served as a small, but central forum for raising and discussing global environmental issues. The Stockholm Conference also served as a precedent for a series of ad hoc United Nations conferences throughout the next decade, each of which had major environmental content. Included were conferences on the law of the sea, population, food, women, water, desertification, human habitation, science and technology, and new and renewable sources of energy. A review of the issues raised at these conferences and how they were dealt with, of the changing perceptions and positions of the actors involved, and of the role of UNEP itself, gives one a good sense of the dynamism of responses to global environmental issues.

An important part of this dynamism relates to the emerging

global environmental values that are examined in chapter 3. An early and basic formulation of these is found in the Declaration on the Human Environment, which was adopted at the Stockholm Conference. This declaration drew upon the theme of mankind living in harmony with nature and expanded it to stress the need and duty of all to act so as to sustain the global ecosystem. More specific goals or values relate to the conservation and preservation of plant and animal species, the conservation of nonrenewable resources, and the maintenance of essential air and water quality. These emerging global environmental values have important implications for other global values such as peace, equity, human rights, and economic development.

How these values complement or conflict with environmental values can be seen as one moves to a consideration of their actual importance or application. Some conflicts relate to the actual distribution of natural and technological resources as was indicated in the example of the Iceland–United Kingdom cod war described earlier. Others relate to two competing legal principles that tend to complicate other issues and values. These are the principles of *national sovereignty* and the *common heritage* of mankind, the international analogues of the ideas of private and public property within states. Chapter 4 seeks to illustrate the interactions of these values and principles through both hypothetical and actual environmental policy case studies. Four studies—oil pollution of the oceans, acid rain, saving the whales and other ocean living species, and wildlife preservation—are then presented to show the complex interactions in developing and implementing environmental policy and to provide materials for student analysis.

The response of the various international actors to global environmental problems has also been shaped by their images of the future. Some of these are based on conventional concepts of progress, themselves largely dependent upon the easy availability of resources and cheap energy. Others are based on fears of either nuclear or environmental catastrophes and seek major reform of the international system. Still others see great social or technological "breakthroughs" radically transforming all levels into some sort of postindustrial society. Chapter 5 seeks to clarify the discussion by distinguishing between probable futures (those based on forecasts from current trends) and possible futures (those futures envisioned by various authors, scientists, businessmen, and policy makers who seek to shape the course of events through their own vision of the future). A number of alternative future scenarios are discussed to show the various value and environmental assumptions that underlie them. The concluding chapter reviews the current state of our global environment and reflects upon the progress made in the decade since Stockholm. Overall, our hope is that students will come to realize that there is still a very great deal to be done to make our global home one in which we can all live in comfort and on a sustainable basis, and that it is both possible and necessary to work in that direction.

2 Environmental Actors

Thousands of people from all over the world converged on Stockholm in June 1972 for the United Nations–sponsored Conference on the Human Environment. These included such well-known figures as His Royal Highness Carl Gustav of the host government Sweden, Kurt Waldheim, Secretary-General of the United Nations, Maurice Strong, the Conference's Secretary-General, Robert McNamara of the World Bank, Shirley Temple Black, Paul Ehrlich, Margaret Mead, Barbara Ward, and other less well-known officials and private persons. What brought them together at Stockholm was a shared interest in global environmental issues.

The UN Conference on the Human Environment was called by the UN General Assembly on the initiative of the Swedish government. The government delegates at the Stockholm Conference discussed and agreed on a number of principles concerning the environment, which are expressed in the Declaration on the Human Environment. They also adopted an Action Plan of 109 items, recommending specific actions to be taken by governments and by international organizations to deal with a wide range of environmental problems. In addition, the Stockholm delegates recommended that the General Assembly establish the UN Environment Program (UNEP), with a small permanent secretariat, supervised by a Governing Council, to administer a modest environment fund for the support of information and education projects.[1]

In this chapter, we will first identify and categorize the relevant actors involved in environmental issues prior to, at, and since the Stockholm Conference. These actors are officials, members, or employees of governments, of international intergovernmental agencies, or nongovernmental associations, and of business entities. Then we will examine the interaction of these various actors as they pursued their goals, expressed their positions, and undertook activities in three different international contexts: first in the Stockholm Conference, next in the UN Environment Program, and finally in the UN-sponsored conferences on problems bearing on the environment that have been held in the decade since Stockholm.

Types of Environmental Actors

As explained in the introduction to this volume, there are several different types of actors involved in any global issue such as the environment. For our purposes, environmental actors are classified into two broad categories—governmental and nongovernmental—according to their status; and into three levels—international, national, and subnational—according to their scope. The nongovernmental category is further subdivided into private nonprofit associations and private profit-making enterprises, and the international (or transnational) level may be further divided into those that are global in scope and those concerned with a particular portion of the earth, such as a region. Table 2.1 illustrates this classification scheme.

Actors in each of these categories have different perspectives on the causes and cures of environmental ills and different capacities to affect the processes of policy making and implementation at various levels. The differences in perspectives and capacities are related to differences in a number of relevant factors—goals, areas of expertise, constituencies, values, etc. These differences are observable not only *between* categories of actors (e.g., nongovernmental associations have a different perspective from governments) but also *within* categories (e.g., the governments of industrialized countries have a different perspective from those of the developing countries). Sometimes the conflicts of interest impede collective action on environmental policies; sometimes it is possible to reconcile the divergent interests and reach some compromise on mutually acceptable action.

Before looking at the Stockholm Conference as an arena in which these different actors expressed their various interests and played various roles in international environmental decision making, a sketch will be provided of goals relating to each type of actor and of the functions they perform with respect to environmental issues.

International Intergovernmental Organizations

International intergovernmental organizations (IGOs), whether global or regional in membership, have perspectives that are at the same

Table 2.1 Types of environmental actors

Levels	Governmental	Nongovernmental	
		Nonprofit associations	Profit-making enterprises
Global	Worldwide inter-governmental organizations such as the UN Food & Agriculture organization	International nongovernmental organizations such as the international Council of Scientific Unions	Multinational businesses such as Mobil Oil, Dole
Regional	Regional intergovernmental organizations such as the European Economic Community	Regional nongovmental organizations such as European League for Economic Cooperation	Enterprises operating in a region such as the Mediterranean basin
National	National governments their officials such as delegates to UN conferences, and agencies such as the US Environmental Protection Agency	National nongovmental organizations such as the American Association for the Advancement of Science	Enterprises operating within a country such as US coal companies
Subnational	State and local government officials and agencies such as air quality control boards, planning commissions	Local citizen groups, individuals	Enterprises operating in a local area such as the utility company operating the Three Mile Island power plant

time both broad and narrow, and roles that are both large and limited. This seeming paradox derives from the nature of those organizations: they are *inter*national, not *supra*national, institutions. Each is set up by a treaty ratified by each of the member governments, whose representatives constitute the policy-making bodies of the organizations. Thus, the combined views of the governments represented, for example, in the UN General Assembly or at international conferences such as Stockholm, may reflect at times a broad transnational perspective on issues and at other times a collection of narrow parochial national perspectives. Further, the IGOs' decision-making capacities are limited by the continued sovereignty of the member governments. Policies adopted, treaties drafted, programs developed—all require for their implementation voluntary acceptance by the member governments. So the IGO role in policy making can be large in the area of policy and program initiation, but may be limited by reliance on governments to enact implementing legislation, to ratify treaties, and to provide financial support for programs.

However, it should be recognized that in addition to government representatives on their policy-making bodies, IGOs also have their own staffs and chief administrative officers who often are not simply passive implementers of the government representatives' decisions. They can be active initiators of proposals, preparers of background documentation and of draft resolutions brought to the government delegates for their consideration, and active participants in the process of building support for environmental action. The Secretary-General of the Stockholm Conference, Maurice Strong, played a very important role during the preparations by influencing Third World governments to view environmental protection as compatible with development, thus averting the threat of reduced Third World participation.

Among the IGOs, not only the UN but also the *specialized agencies* of the UN system and various regional organizations are actively involved in environmental matters. The specialized agencies, global in membership but limited in purpose to some specific area of international cooperation such as health, food and agriculture, or weather forecasting, have generally exhibited a "sectoral" perspective, approaching environmental issues from the viewpoint of each agency's own area of responsibility. Agencies such as the United Nations Educational, Scientific, and Cultural Organization (UNESCO) and the Food and Agriculture Organization (FAO) had in fact undertaken environmental programs even before the United Nations turned its direct attention to the environment at Stockholm. UNESCO had initiated scientific and educational programs such as the Man and the Biosphere program and had cooperated with other international agencies in organizing the International Geophysical Year. The FAO had been engaged in programs to increase food production and to improve forest and soil management practices. What each specialized agency may lack in perception of the "big picture," it makes up in expertise in its own area of competence.

Of course, each specialized agency is eager to increase its own responsibilities, staff, and budget, and to maintain its autonomy in program development and execution. This desire has led to "turf battles," over which IGOs will have what responsibilities with respect to the environment. Since some of the specialized agencies had already carved out for themselves a piece of the "turf" even before the Stockholm Conference, they successfully resisted the proposal to create a new specialized IGO for the environment. Since Stockholm, they have sought to define narrowly the UN Environment Program's intended role as coordinator of all the environment activities of the UN family agencies.

Regional organizations, initially involving the industrialized countries of Europe, but now including some in other areas as well, have also made efforts toward environmental cooperation. As with the specialized agencies, the regional IGOs have encountered the problem of "sectoralism," that is, how to integrate the various sectors into a combined attack on environmental problems, which are, as we know, interdisciplinary in nature. A further problem for these organizations has been how to integrate their environmental activities with the other economic, social, and political issues with which they are primarily concerned.

Each of the principal European organizations has devised a different mechanism for intraorganizational cooperation and coordination and for the development of a comprehensive environmental program. The Council of Europe has since 1962 had an intergovernmental Committee for the Conservation of Nature and Natural Resources, but its work is limited to the conservation of wildlife and natural habitats. The Organization for Economic Cooperation and Development, which includes Western Europe, Canada, the United States, Japan, Australia, and New Zealand, created an Environment Committee in 1970 with a multidisciplinary mandate to consider technical and scientific developments as well as their financial, economic, and social implications. In 1971, the Economic Commission for Europe (ECE), a regional commission of the UN that encompasses both Eastern and Western European countries, appointed a committee of Senior Advisers on Environmental Problems. They work on problems requiring a multidisciplinary approach and seek to complement the environmental work done by the sectoral ECE committees on the chemical industry, coal, electric power, gas, steel, timber, water, housing, building, and planning. The European Economic Community (often called the Common Market) created the Environment and Consumer Protection Service in 1973. It is responsible for the coordination of environmental activities, for the elaboration of a community environment program, and for the execution of projects that do not fall within the competence of the Directorates for Transport, Agriculture, Industry, Research, and Social Affairs. Also in 1973, the Eastern European socialist countries' Council for Mutual Economic Assistance established the Council for Environmental Protection and Improvement, with six intergovernmental councils and six coordination centers to deal with ecosystems and landscapes, air pollution, solid waste management, environmental health, and the economic, legal, and educational aspects of environmental protection. Even NATO joined the environment bandwagon, creating the Committee on the Challenges of Modern Society to consider environmental problems.

National Governments

National governments are, of course, the single most important category of environmental actors, because both international and national action requires their approval in policy making and implementation. National governments ratify (or fail to ratify) and comply (or fail to comply) with international treaties such as the 1973 Convention for the Prevention of Pollution from Ships. National governments also contribute to IGO projects such as the UN Environment Fund and they may fully or partially implement such internationally developed programs as the 1972 Stockholm Action Plan and the 1980 World Conservation Strategy. National governments enact legislation such as the U.S. Clean Air Act, they create ministries for the environment or other agencies such as the U.S. Environmental Protection Agency, and they provide tax and other incentives to industries, local governments, and individuals, which affect the environment directly or indirectly.

Government perspectives on environmental issues vary from country to country, but a concern common to all is the preservation of national sovereignty. That goal leads governments to resist any encroachment on national autonomy in decision making and to insist that international environmental measures be based upon international cooperation, with legal obligations limited to those to which states have directly agreed. Beyond this shared view on sovereign prerogatives, states' perspectives vary widely depending on both economic and political factors.

The industrialized market-economy countries, such as the United States, were the first to put environmental protection on the public policy agenda, and their taxpayers and consumers are financially better able to bear the costs than are those of developing countries. However, the particular economic interests of each industrialized country affect its support for or resistance to particular environmental measures. Japan, for example, dissented strongly from the 1982 decision of the International Whaling Commission to phase out all commercial whaling by 1985. Japan's dissent is attributable to the Japanese use of whale meat for food and for the individuals who make their livelihood from whaling. The market-economy countries' perspectives are also influenced by their commitment to the principle and practice of private property. The political act of balancing private owners' rights to use their property for their individual benefit with society's right to regulate such use in the public interest has a continuing significance not just for nations, but for all levels of government.

The centrally planned industrialized countries such as the Soviet Union have, in principle, no such public-private issues to contend with, but they are not free from competition among various sectors of industry and agriculture, each with its goals of maximizing production and minimizing costs. In addition to the economic issues, the allocation of human, material, and financial resources involves political considerations, such as the relative emphasis to be given to national defense, industrial and agricultural expansion, housing, environmental health and safety, etc. Further, the socialist countries' commitment to the goal of improving the material conditions of all their citizens presumes an ever-increasing output of goods and services. Socialist visions of an inexhaustible cornucopia of resources are resistant (as in nonsocialist countries as well) to the message that there are material limits to growth imposed by the earth's carrying capacity.

The developing countries, such as India, have perspectives on ecological matters that are screened through the twin objectives of economic development and the assertion of national independence. The developing countries at first reacted negatively to the industrialized countries' sounding of the environmental alarm and their urging that the developing countries not repeat *their* errors. Third World countries interpreted those messages as an ill-concealed attempt to retard development in the Third World and to perpetuate a self-serving neocolonial paternalism. This initial suspicion has been overcome to some extent. Governments in many developing countries are increasingly aware that development projects undertaken without regard for their adverse environmental effects may bring benefits in the short run, but at unacceptably high long-term costs. Some governments, however, still feel compelled by economic circumstances to open their countries to manufacturing and other foreign enterprises in flight from pollution regulations in their home countries. Yet the concept of development that is sensitive to ecological impact—"eco-development"—is gaining support; its message is that development must take into account the particular environmental and social conditions in each country or region, must apply technology appropriate to those conditions, and must respect local sociocultural patterns, rather than destroy them by the imposition of alien patterns of organization.

Third World countries have also come to see environment-development issues in the context of their larger goal of building a New International Economic Order by restructuring current relationships in a way that will narrow the gap in living conditions between the rich industrialized countries of the North and the poor developing countries of the South. The South asserts that poverty itself is a major cause of environmental degradation and that the North, having benefitted from exploitation of the natural and human resources of the South, has an obligation to assist them in overcoming that poverty. Some of the industrialized countries are reluctant to accede to Third World requests for expanded technological and financial aid, such as the proposal of the 1981 Conference on New and Renewable Sources of Energy to establish a new fund to help the poorer countries develop energy sources. The proposal was resisted by the United States and other Western countries.

While no single characterization of Third World perspectives can accurately portray the views of such a large and diverse group of states, there is among them a rather widespread reluctance to subordinate economic needs to environmental protection, a rather widespread conviction that the costs must at least in part be borne by the rich countries, and that external restrictions on their control of their own natural and financial resources must be resisted. Third World governments, like their counterparts in the First and Second Worlds, also face conflicting internal pressures on environmental policy making. A case in point is the government of Kenya, which is pressured by its national tourist industry (as well as by international conservation groups) to preserve the game parks and the wildlife that attract foreign tourists and are a major source of national income. At the same time it is pressured to open up the parks to farmers and pastoral people and to allow killing of wildlife for food in order to meet the needs of a rapidly growing population. All governments confront conflicting demands, but for the poorer countries the margin of choice is even more constricted than it is for affluent societies.

Subnational Governments

While national governments are the prime wielders of power over environmental policies, local and state governments are not without significance. They make decisions on environmental matters within their jurisdiction and they take action that stimulates national efforts. In the former category is the considerable environmental impact of such city and county decisions as those on zoning, building permits, road construction, park acquisition, and flood control, to mention a few. And state and local governments often advocate national action on environmental problems that they discover are beyond their legal or financial capacity to solve (the states, for example, have pressed the federal government in the United States for policy initiatives and legislation to control acid rain).

Occasionally local or state governments may resist efforts of national governments to impose national standards. An illustration of this is the resistance of some states to the national 55 mph speed limit, instituted to conserve gasoline by Congress through its authority over the Interstate Highway System. Another case, in which a state government demanded more stringent regulation within its borders than Congress had mandated by federal law, is California's establishment of emission control requirements on automobiles sold or registered in that state.

Subnational government entities are usually not visible in international settings such as the Stockholm Conference or the UN Environment Program; it is clear, however, that many of their daily activities have substantial effects on the natural and human environment even though their perspectives may be geographically limited to their own areas of governance.

Nongovernmental Associations

Private nonprofit associations, from local citizens groups to national conservation organizations, to international scientific or general interest organizations, constitute one of two categories of influential nongovernmental actors, the other being business enterprises. For NGOS, the great variety of interests they represent makes it impossible to generalize about their perspectives and goals. Some idea of the range of those interests is conveyed by the kinds of organizations (most of them international NGOS) that were involved in the Stockholm Conference. Among the 237 such associations, 21 different fields of interest were represented, the most numerous ones being social welfare, religion and ethics, education and youth, international relations, law and administration, science, professions and employers, technology, health and medicine, and commerce and industry.[2] The three international NGOS that were most visibly involved before and during the Stockholm meeting were the International Council of Scientific Unions (ICSU), the International Union for Conservation of Nature and Natural Resources (IUCN), and the Friends of the Earth, all of which have also been closely involved since Stockholm with UNEP and in other environmental programs.

National associations are similarly diverse in their separate fields of interest and many of them have organized into international federations. The U.S. Chamber of Commerce, for example, is a member of the International Chamber of Commerce, and is itself a federation of state and local chambers of commerce in the U.S. Thus many NGOS have multilevel affiliations, which give them access to governmental bodies at different levels. Local citizen groups, of course, are sometimes single issue and ad hoc, mobilized for such purposes as cleaning a toxic waste dump site in their neighborhood or resisting the presence of an oil refinery or nuclear power station in their town or region. Once the goal is achieved, however, the group is likely to disband.

The activities of NGOS are not confined to the advocacy role of bringing pressure to bear on local, state, and national officials or on international organizations (although the pressure group function is one that is widely accepted, at least in countries with pluralist political systems where interest groups are most numerous). Another role, typical of the "expert" scientific and technical NGOS, is to provide specialized knowledge to governmental and intergovernmental decision makers. ICSU, for example, was asked before Stockholm to prepare a plan for a global environmental monitoring system, which has now come into being and is known as "Earthwatch." Scientific NGOS often collaborate with IGOS in environmental assessment projects, data collection and evaluation, and in program planning. The IUCN worked with UNEP, the World Wildlife Fund and selected governments to develop the World Conservation Strategy, which was launched in 1980 with a worldwide publicity effort. Nonexpert NGOs in fields such as religion, youth, and international relations often conceive of their function as being that of transmitters of information to their members or to the general public on ecological problems and on what is being done or ought to be done about those problems. And finally some NGOs carry out their own environmental projects, raising money and enlisting the services of their members in such activities as tree planting, recycling, or the acquisition of parks and ecological preserves.

Business Enterprises

The paramount purpose of business enterprises is to seek profits and growth. This profit-making objective defines their perspective on environmental issues and leads them to resist measures that inhibit the pursuit of that goal. It is not that the managers, workers, and owners of enterprises are insensitive to environmental concerns, but rather that they accord higher priority to profitable management, to jobs, to dividends. In California, for example, container manufacturers and retail stores successfully opposed a 1982 ballot measure to require deposits on beverage cans and bottles; their opposition to the measure was presumably not because they like littered highways or object to conserving energy or to recycling aluminum, but rather because returnable containers would involve additional handling costs for retail stores and reduce demand for the products of container manufacturers. Business resistance to environmental measures that require changing production methods, curtailing operations, installing new equipment or making other adjustments is reduced when the enterprise is able to pass the additional costs on to the consumer (such as in pollution-control devices in automobiles) or when government subsidizes the cost either directly or through tax benefits to the company. Sometimes business interests conflict with each other over environmental measures. In scenic coastal areas, for example, local hotels and other enterprises dependent on attracting tourists find themselves in opposition to offshore oil drilling, which risks polluting the beaches and waters, while national and international oil companies pressure the government to grant leases for such drilling.

No listing of environmental actors would be complete without mention of one important actor—each of us. As consumers, as employees, as recreation seekers, as citizens, we affect our immediate environment and have an impact on the global situation. That impact is felt when we conserve gasoline and reduce air pollution by proper auto maintenance, when we install solar collectors on our roofs, when we recycle cans and paper, when we treat wilderness areas with care, when we support environmental legislation and other forms of community action. Our attitudes toward the "commons" we share with all humans and other living creatures, the values we hold, the actions we take, all count not only for us now but for the future.

Environmental Actors and the Stockholm Conference

At the UN Conference on the Human Environment, the representatives of national governments were the primary conference participants as 113 nation-states, fewer than the total UN membership, sent delegations. This less-than-universal participation was due to the Soviet and Eastern European countries' decision not to attend because East Germany (which, like West Germany, is not a member of the UN) had not been invited to Stockholm, although West Germany had been. However, the Soviet bloc countries' nonparticipation at Stockholm did not mean that they were opposed to environmental cooperation; on the contrary, they were actively involved in preparations for the conference and have supported not only UN environmental programs but also a number of bilateral programs (with the United States). In addition to the government delegates, the Stockholm Conference also officially involved members of the UN Secretariat and representatives of the specialized IGOs such as the World Bank, whose financial support for development projects involves it in environmental issues; the FAO and UNESCO, some of whose environmental projects were mentioned above; and regional IGOs, such as the Council of Europe.

Present in Stockholm and participating in a wide range of activities parallel to the UN conference were members of more than five hundred nongovernmental organizations representing diverse interests. Some were global conservation groups such as the IUCN. Others, such as the Sierra Club, represented a more limited national constituency. Scientific groups such as ICSU were there, as were many religious, youth, business, women's, and other groups with international, national, and local memberships. Commercial enterprises were less visible as participants at Stockholm, although their interests were represented through business NGOs, most notably the International Chamber of Commerce, and through their influence on national governmental delegations as well.

Controversies arose among these various actors before and during the conference, reflecting the competing objectives pursued by the different actors. In some instances, the controversy was not over *whether* to protect the environment, but over *how* it should be done and what priority should be assigned to *which* environmental problems. These different objectives were pursued by various means. For government delegates, the means included diplomacy, debate, and legislative tactics—introducing, amending, and passing (or defeating) conference resolutions such as the Declaration on the Environment, the Action Plan, and recommendations for future UN institutional mechanisms. For the nongovernmental representatives, who were not official conference decision makers, views were expressed and objectives promoted by attempts to reach the national delegates and the Conference Secretariat. Some of the international NGOs, those that had been accorded "consultative status" by the UN, were permitted to address sessions of the conference; they and nonrecognized groups participated in a "parallel" conference called the Environment Forum, and in a number of other special events designed to bring particular issues (such as the threatened extinction of certain varieties of whales) to the attention of the government delegates and the public.

Contributions of the Stockholm Conference

The Stockholm Conference achieved several important conceptual advances. Most of these advances emanated from the Founex meeting, convened in 1971 by Conference Secretary-General, Maurice Strong, as a first attempt to define the relationship between environment and development and to build a framework for reconciliation of perspectives between the North and the South. The inseparability and compatibility of environment and development were accepted. Environmental problems in Third World countries were defined to include the "pollution of poverty" and the negative consequences that accompany the process of growth. Thus, a broad approach was taken to development, one going beyond the usual focus on economic growth as measured simply by increases in the Gross National Product (GNP). The redefinition of development goals and objectives to include ecological, cultural, and social factors was a significant event.

Other important conceptual contributions of the Stockholm Conference can be identified.

1 The conference was the first intergovernmental meeting to challenge the unquestioned acceptance of growth as the indicator of economic development and technological solutions as the means to achieve it.
2 The need for a change in institutional methods and arrangements was identified. Traditional relationships between sectors based on coordination among separate units were shown to be inadequate to deal with environmental problems. A transdisciplinary approach that cuts through and links all areas was suggested.
3 Stockholm gave impetus to a process of reflection and discussion that challenges some of the basic tenets of modern industrial society.
4 One of the biggest achievements of the conference was its avoidance of a stalemate between the "polluter must pay" approach and the "no growth" approach, which was a prominent environmental posture at that time. In fact, the conference took a third route; it oriented the discussion toward a critique of the process of development.

The above achievements are enough to rate the conference a success. Their significance should, however, not keep us from realizing that the conference did not deal with some of the more radical and fundamental issues implicit in the global environment. The Declaration on the Human Environment redefined global objectives and called for a broadening of the scope of international cooperation, but there was no analysis of some of the more fundamental problems of the international economic and political system. Basic structural issues were kept off the agenda. Missing were discussions of such topics as land use, exploitation and distribution of natural resources, alternative patterns of resource use, the social and ecological implications of the Green Revolution, and the structure of industrial economies. The fact that countries were still preoccupied with trying to define what environment means suggested that it was premature to pursue the implications of the link between environment and the international system and domestic institutions.

With hindsight, the Declarations and Action Plan are in some ways disappointing, not with respect to what was included but with respect to what was omitted. Yet given the basic interests of the North and South, agreement was reached on a higher common denominator than would have been expected. Confrontation was avoided and a basis for dialogue was set that would help sustain this willingness to avoid a North-South split. Interests differed, but the dialogue remained. The primary concern of the South was to gain assistance for accelerated development and to find ways to avoid the possible adverse economic effects of environmental measures. The industrialized countries were concerned with disequilibrium in international trade and investment patterns, and in pollution control. Priorities and interests were different, although complementary. With regard to industrialization, the South's desire for industrial development at any cost suited the North's desire to export its most polluting industries. Acceptance of the principles of protection of the global commons as a basic human right was significant, but abstract enough to hide the fact that there was no fundamental understanding between the North and South on either causes or solutions, no firm commitment to carry out the recommendations, and no agreement on the type of international order that would permit the impressive verbal formulations to be translated into concrete action.

The New Environmental Actor: The UN Environment Program

At Stockholm there was considerable debate over whether to create a new intergovernmental organization to deal with environmental problems. The specialized agencies of the UN system in particular were opposed, each preferring instead to be given responsibility for environmental concerns in its own sector. For example, the FAO would be responsible for environmental issues relating to agriculture and forestry, the World Health Organization for water, sanitation, diseases, and health, and UNESCO for the study of scientific aspects of the environment and for developing materials for environmental education.

The position of the specialized agencies plus concern by the industrial countries about high costs led to a decision not to establish a new specialized agency for the environment. Instead the General Assembly in the fall of 1972 created the UN Environment Program (UNEP), a small unit of the UN Secretariat, directed by a Governing Council of fifty-four member governments. The General Assembly accepted the invitation of the government of Kenya to locate the new program in its capital city, Nairobi, where a large conference center honoring Kenya's first president, Jomo Kenyatta, was then under construction. The decision to put UNEP's headquarters in Africa had symbolic significance as the first UN unit (other than regional economic commissions) to be located outside Europe and North America. Headquartering UNEP in Africa was seen as a political recognition of the growing importance of the Third World in international affairs and also as an assertion that environmental issues are as important to the developing South as they are to the industrialized North. These considerations overrode the reservations of some governments and international organization officials who feared that Nairobi's geographic distance from UN headquarters, from existing centers of scientific research and data, and from the principal environmentally concerned NGOs might hinder UNEP's capacity to carry out the responsibilities given to it.

UNEP's mandate and its institutional and financial arrangements reveal a series of assumptions about the nature of environmental problems and about UNEP's proper place in the UN system.

1 Environment is not another sector, but an interdisciplinary problem that must be considered in all sectoral activities.
2 No new institution should be created with the sole responsibility for implementing environmental projects. Each UN agency should be responsible for the environmental aspects of its activities within its own specialized field of competence.
3 A small entity (UNEP) should be created that would: (*a*) perform a "think-tank" function by producing innovative approaches and inspiring, motivating, and persuading already existing bodies to integrate the environmental perspective into their activities; (*b*) be a focal point for coordination of environmental activities in the UN system; (*c*) concentrate its professional competence in multisectoral areas that transcend the fields of competence of the sectoral organizations, such as monitoring, oceans, human settlements, and deserts; (*d*) be nonoperational or, in other words, not carry out projects itself, but delegate or encourage governments, member organizations of the UN and nongovernmental organizations to do so.
4 In order to carry out its coordination role this new entity would have to "meddle" in the affairs of others, but should not divide responsibilities; rather it should promote interaction within the UN system and between the UN, other IGOs, governments, and NGOs, with the purpose of reaching joint definitions of goals and programs.

UNEP was thus created as a mechanism whereby a small catalytic group could motivate and stimulate the interaction of all the actors in such a way as to produce a common, flexible, and evolutionary global context for national policies. To accomplish this objective, UNEP was given a mandate to influence change. For the UN system its design was very innovative in the following ways.

1. Politically, its location offered an opportunity to bridge North-South and East-West conflicts, to demonstrate a common bond, and to work toward positive common goals in an interdependent world.
2. Philosophically, the issues it was to confront and contemplate raised basic questions about the survival of mankind and the planet.
3. Conceptually, it was expected to challenge the present order and suggest innovative and structural approaches to ensure a compatible relationship between environment and development.
4. Functionally, it was given a unique charge to coordinate the environmental activities of the UN by the integration of environmental dimensions into sectoral and development activities.
5. Organizationally, it was given responsibility to set up a process by which all concerned actors could participate in the elaboration of its policies and in their implementation.

With this set of original expectations for UNEP in mind, let us now look at how UNEP has met those expectations in the decade since its founding.

UNEP's Contributions

Just as the Stockholm Conference made important contributions to the way governments and other actors think about the environment, so also has UNEP made substantial contributions, particularly in the evolution and application of the concept of "ecodevelopment." First, its innovative interpretation of the relationship between environment and development avoids the pitfall of viewing environmental concerns as an obstacle to development. This has been a major factor in obtaining the support of Third World countries and has helped avert a North-South confrontation. Second, UNEP has placed the whole issue in a structural setting. As UNEP approaches it, the problem is not how to prevent or end pollution, but how to give an environmentally sound direction to development strategies. In advancing and promoting the concept of sustainable growth, UNEP became the first UN body to set out an alternative development strategy based on the spirit of self-reliance and comprehensive long-term structural reform. Furthermore, it has attempted, in its emphasis on management and planning, to make eco-development an action-oriented concept and to avoid purely speculative discussions.

UNEP's work on eco-development, sustainable growth, and structural reform are in some ways parallel to the Third World's larger demands for a New International Economic Order (NIEO), while in other ways rather different. Although the Stockholm Conference contained some of the seeds for NIEO, it took the oil price increases after the 1973 embargo and the debates in the Sixth Special Session of the UN General Assembly (devoted to raw materials and development) to produce the formal Declaration on the Establishment of a NIEO (1974). Both NIEO and eco-development are concerned with global poverty and how to deal with it. The NIEO calls primarily for an international redistribution of wealth. This is to be achieved through changes in the basic terms of trade between North and South, increasing credit and developmental assistance to the South, and a reduction of barriers to the free flow of goods, commodities, and technologies. Eco-development, while not denying the utility of these initiatives, seeks to reduce poverty by encouraging environmentally sound development, something that often requires internal structural change rather than international change alone. NIEO tends to focus more on industrial and economic development; eco-development stresses more the need to focus on rural areas, on agriculture, and on resource limits in order to develop programs for sustainable growth. The NIEO conception is still largely one of technological optimism and unlimited economic growth. UNEP, through its work on eco-development and sustainable growth, has cautiously raised questions about the costs of growth and possible limits. As in other matters, it has had to walk a tightrope in doing so. In raising and dealing with such issues, UNEP has been able to draw upon a series of emerging global issues and conferences to broaden an awareness of questions of global limits and how they relate to the development choices of both the North and the South.

UN Conferences on Global Issues

The UN Conference on the Human Environment at Stockholm was the first of a series of UN-sponsored world conferences on global issues, most of which have some connection—immediate or more remote—to the environment. In the decade since Stockholm, there have been conferences on population (Bucharest, 1974), food (Rome, 1974), women (Mexico City, 1975), human settlements (Vancouver, 1976), water (Mar del Plata, 1977), desertification (Nairobi, 1977), science and technology for development (Vienna, 1979), new and renewable sources of energy (Nairobi, 1981), and the UN Law of the Sea Conference, which met in various places intermittently from 1958 until its recent conclusion.[3] The 1982 special session of UNEP's Governing Council brought the meetings full circle back to environmental issues, as the council concentrated on the global situation "Ten Years After Stockholm."

Stockholm provided the model in terms of preparations, participants, and outputs for most of these conferences. As at Stockholm, government officials and experts have prepared for the conferences by assembling position papers and reports on the issues for discussion at the meetings, and the intergovernmental preparatory committees have hammered out draft declarations and action plans. As at Stockholm, the participants have included representatives of UN members and other governments, UN Secretariat personnel, observers from interested specialized and regional agencies, and persons representing a wide variety of international, national, and local NGOs. The outputs of the conferences have, like Stockholm, customarily been a set of principles expressing the governments' shared views of the problem under consideration and an action plan listing what ought to be done by governments and other actors. Unlike the other conferences, the Law of the Sea Conference produced a lengthy multilateral treaty for ratification by states; its effect, when in force, will be to create legal obligations on nation-states to comply with its provisions, an obligation that does not apply to the other conferences' action plans.

The conferences are thought to have served the function of directing attention and keeping it directed toward the pressing global issues of our times. This global "consciousness-raising" has affected the government delegates and others who attended the conferences and who established contacts with each other that have become durable communication networks. It has also educated the officials and private persons who have been involved in the preparation of position papers, expert reports, and other conference documentation and have been the audience for information on conference issues circulated through the mass media, though government, NGO, and IGO publications, and through NGO-sponsored programs. In the words of one observer, the conferences "may well be seen as historic. These giant sensitivity training sessions . . . are designed to gather a large crowd from all over, and raise the world's attention level for an important part of the interdependent whole."[4]

All of the conferences have dealt with subjects that are connected to the environment. Population growth, for example, clearly has environmental effects, as the increased demand for food, shelter, and employment places accelerating pressure on natural resources. Food to meet the needs of a growing population requires increasing production by such means as use of fertilizers and pesticides, irrigation, and expanding cultivation into forested lands. The social roles of women as workers and as childbearers are related to economic production and population growth. Water for human consumption and for agriculture and industry affects public health and population growth, as well as economic development and human settlement patterns. The application of science and technology has major impacts on both the physical and human environments in which development occurs. The conferences with the greatest direct relevance to the environment are those on human settlements, desertification, energy, the law of the sea, and of course the special tenth anniversary session of UNEP's Governing Council.

The Human Settlements Conference

The Human Settlements Conference, known as Habitat, was a direct outcome of Stockholm, where it became clear to the delegates that many of the issues involving the condition of the natural environment are inextricably linked to the man-made environments of cities, towns, and rural settlements. It was agreed at Stockholm that a follow-up conference would be held and the Canadian government offered to host that event in Vancouver in 1976. Governments put a major effort into preparations for Habitat, including lengthy written inventories of their current policies and programs concerning land use, housing, sanitation, transportation, health, education, and the whole range of social and economic services to urban and rural populations. NGOs also prepared many reports for distribution at the conference and, as at Stockholm, put together a parallel conference, the Habitat Forum, which offered lectures, symposia, films, entertainment and even an exhibit-demonstration of low-cost, owner-built, environmentally sound housing units for visitors to see.

The customary statement of principles, the Vancouver Declaration on Human Settlements, was adopted, along with an action plan addressed mainly to national governments, stressing the need for planning land use and for action to meet such basic human needs as satisfactory housing, food, clean water, education, and jobs. A final outcome was the recommendation that the UN General Assembly establish permanent UN machinery for human settlements. As a result, the Centre on Human Settlements, also called Habitat, was created in 1978, replacing the Economic and Social Council's Committee on Housing, Building, and Planning. The Habitat section of the secretariat is, like UNEP, located in Nairobi and is supervised by an intergovernmental Commission on Human Settlements.

The Desertification Conference

UNEP played a major role in the UN Conference on Desertification held in Nairobi in 1977.[5] The severe drought in the Sahelian region of Africa had directed world attention to the continuous expansion of deserts in Africa and elsewhere. The General Assembly accordingly called for an international conference to consider the causes of and remedies for desertification. UNEP's Executive Director served as Secretary-General for the conference, and its Governing Council as the intergovernmental preparatory committee. The UNEP Secretariat organized regional preparatory meetings, consulted with other UN agencies such as the World Meteorological Organization, FAO, UNESCO and the UN Development Program, and collected information from scientists and technical experts, which was incorporated into the conference documentation.

A Plan of Action to Combat Desertification was adopted and UNEP was given the responsibility to follow up on the implementation of the plan. Coordination of activities to halt the advance of deserts and reclaim desert lands is handled by a unit in the Secretariat and through an interagency Consultative Group for Desertification Control, in which UNEP, the agencies mentioned above, and other UN bodies participate. Financing for desertification control projects has been very limited, and progress thus far has not been notable.

The Energy Conference

The oil embargo of 1973 and subsequent meteoric rise in oil prices were prime factors in directing attention to the problem of the world's heavy reliance on energy derived from hydrocarbons (petroleum and gas), which are finite and often produce pollutants as well. The impact of price increases was felt worldwide, but especially acutely in those developing countries that must import oil for industry, agriculture, and household uses. The urgent need to develop and use other cheaper and renewable sources of energy (such as solar, geothermal, wind, wave and tidal power, peat, biomass conversion) was recognized and a conference, the UN Conference on New and Renewable Sources of Energy, was held in Nairobi in 1981.

The program of action adopted there committed participating countries to make the "energy transition from the present international economy based primarily on hydrocarbons to one based increasingly on new and renewable sources of energy in a manner which, consistent with the needs and options of individual countries, is socially equitable, economically and technically viable and environmentally suitable."[6] National actions are to include assessment of energy needs and sources, preparation of energy-policy plans, support for research, development, and training, and programs to encourage expanded use of alternative energy sources. International cooperation is to be directed particularly toward the needs of developing countries by concerted action in the areas of energy assessment and planning, research and development, transfer and adaptation of technologies, information exchange, and education and training.

Despite this ambitious plan, and despite the wish of the developing countries that a new UN institution be established with sufficient funds to help them develop and use alternative energy sources, the industrialized countries successfully opposed any additional UN staff for that purpose, and the United States opposed the idea of creating a special world energy fund. Other industrialized countries such as the Common Market members, the Scandinavian countries, and Canada, however, supported that idea. For the present, the implementation of the energy action plan is being carried out through various separate local, national, and regional means, coordinated by an interdepartmental working group with the UN.

The Law of the Sea Conference

The prolonged deliberations over the international laws governing the oceans and seabed beyond national jurisdiction have occurred largely as a result of technological developments that created threats to the marine environment: opening the seabed to oil drilling and possibly to other mineral exploitation, changing fishing techniques to the point that some species of fish are threatened with extinction, increasing marine pollution from ships and shore sources, and expanding national claims of jurisdiction over the continental shelf, territorial seas, and other areas, which have led to clashes of sovereignty.

The comprehensive Law of the Sea (LOS) Convention, finally adopted by the Conference in 1982, is an effort to codify existing international law and to expand it to cover new issues that have arisen particularly since World War II. Recognition is given to the principle that the resources of the seabed and ocean floor beyond territorial jurisdiction are the "common heritage" of mankind, to be accessible to all nation-states. Those resources are to be managed by an International Seabed Authority with power to license companies to mine the seabed and to engage in mining itself. The income from those mining activities is then to be used to help poorer countries, enabling them to share in the ocean's wealth. The United States voted against the treaty, objecting in particular to the seabed mining provisions and preferring instead an agreement among potential mining countries, all of which are developed.

The somewhat less controversial sections of the LOS convention concerning marine pollution, marine research, fisheries, maritime boundaries, and other topics call for UNEP, FAO, the International Maritime Organization (IMO) (formerly called the Inter-Governmental Maritime Consultative Organization or IMCO), and UNESCO's Inter-Governmental Oceanographic Commission to identify experts to participate in the arbitration procedures provided for in the treaty. UNEP and IMO have the responsibility of calling regional or world conferences to draw up rules and standards for the protection of the marine environment from pollution.

Ten Years After Stockholm

In 1982 the UNEP Governing Council marked the tenth anniversary of the Stockholm Conference by holding a special session devoted to a review of the measures taken during the decade to carry out the 1972 Action Plan and to point out what future measures need to be taken. At the same time the NGOs also held a Symposium on Environment and the Future in Nairobi. Some important advances have been made since Stockholm. The number of states with national environmental departments or other agencies for the environment has increased from ten in 1970 to nearly one hundred in 1980, and the number of NGOs with environmental interests has increased sixfold. UNEP's assessment programs, most notably the Global Environmental Monitoring System, are in operation. In the field of international environmental management, UNEP's Regional Seas Program has made considerable progress in enlisting government cooperation to control pollution in the Mediterranean and other seas. The final declaration adopted at the 1982 Governing Council session also points out that advances have been made in our awareness and perception of environmental problems; there is wide recognition of "the need for environmental management and assessment given the intimate and complex relationship between environment, development, population, and resources."[7]

However, major problems remain, "due mainly to inadequate foresight and understanding of the long-term benefits of environmental protection, to inadequate coordination of approaches and efforts, and to unavailability and inequitable distribution of resources."[8] Among the alarming and increasing problems are those already introduced: deforestation, soil and water degradation, desertification, environmentally related diseases, atmospheric changes such as in the ozone layer, carbon dioxide buildups, and acid rain, pollution of the seas and inland waters, hazardous wastes, and extinction of animal and plant species.

What is needed in particular, according to the council, are efforts to use natural resources in an environmentally sound way, to modernize traditional pastoral systems, to promote innovations in recycling, conservation, resource substitution, and to develop alternative energy sources. Efforts by developing countries at the 1982 Governing Council to expand UNEP's management activities to promote sustainable development were resisted by other member countries, which are reluctant to increase staff and contributions to the Environment Fund. While some countries—among them Denmark, Japan, Libya, the Netherlands, Sweden, the United Kingdom, and Saudi Arabia—did increase their contributions, the United States reduced its share from $10 million to $7.8 million. The role of the U.S. delegation at Nairobi was seen by observers such as Maurice Strong as being "in stark contrast to its leadership role in Stockholm."[9]

The Declaration concluded with a reaffirmation of commitment to the Stockholm Declaration and Action Plan and called for more national efforts and international cooperation, urging "all Governments and peoples of the world to discharge their historical responsibility, collectively and individually, to ensure that our small planet is passed over to future generations in a condition which guarantees a life in human dignity for all."[10]

In this chapter we have identified the major environmental actors at international, national, and subnational levels and shown how they have been involved in environmental policy making and action programs. Their perspectives on environmental issues have been sketched. The next chapter examines the emerging environmental values that the various actors are coming to accept and pursue.

3 Environmental Values

Widespread concern over the impact of human activity on the ecosystem of the planet has emerged into a growing commitment to environmental values, which has been especially noticeable over the past two decades. Contemporary environmentalism first became established in the industrial world, where air and water pollution and the ravages of resource development on the landscape were the most evident. Over the past two decades, public opinion surveys have consistently revealed strong support for environmental values even during the economic downturns of recent years. In the developing or Third World there is also now a growing awareness of the seriousness of ecological problems and the shortsightedness of rushing headlong into development programs without taking into account the long-term costs of environmental degradation. In the international realm, the convening of the Stockholm Conference of 1972 and the establishment of the United Nations Environment Program were manifestations of this growing sense of environmental awareness. One should not, however, infer from these trends that the environment has become one of the most important priorities of the international community. Environmental issues have yet to command as much attention in international circles as political conflicts such as the Middle East, arms races and arms control, economic development, and human rights.[1]

There are also different degrees of environmental concern. Few would argue that environmental quality is not a desirable condition. But for many, maintaining the quality of the natural environment is only one of many desired outcomes, and a secondary one at that. Thus, they are willing to compromise environmental goals that come into conflict with other aspirations, especially economic ones, such as employment. At the other extreme are those for whom the environment is the overriding concern. To them, other goals are to be pursued only in ways that are environmentally responsible. Moreover, achieving environmental objectives may require major sacrifices, including fundamental changes in life-style, such as a reduced reliance on the private automobile.

Underlying the discussions of environmental values is the philosophical issue of why we should be concerned about preserving our natural surroundings. The most common response is that it is in our best interests as human beings to do so. Air pollution, for example, is believed to be a cause of certain types of cancer and has been conclusively linked to respiratory problems. Erosion of topsoil due to careless farming practices reduces the productivity of agricultural land and thus the availability of food. Concern may not be limited to the welfare of the current generation, but also to its descendents whose quality of life may be diminished by today's irresponsible practices, such as the profligate use of scarce, nonrenewable resources and the irreversible destruction of species of plants and animals with useful genetic traits.

These arguments, which are based on humanistic grounds, have been challenged by a smaller number of devoted environmentalists who take the view that man, as an integral part of the natural world, has no overriding moral claim that justifies manipulating it for his own purposes. Advancing what he describes as the "Noah Principle," David Ehrenfeld suggests that nonhuman communities and species "should be conserved because they exist and because their existence is itself but the present expression of a continuing historical process of immense antiquity and majesty. Long-standing existence in Nature is deemed to carry with it the unimpeachable right to continued existence."[2] On many practical matters pertaining to the environment, those subscribing to the Noah Principle will be in agreement with the humanists. However, there are a few issues on which the two schools of thought sharply diverge, as for example whether the smallpox virus should be preserved as an endangered species. Humanists would be inclined to eradicate it on grounds that it causes many great suffering, whereas the Noah Principle would suggest that the virus has as much a right to exist as any species.

Having set the broad philosophical context, let us turn to an examination of the more prominent values and goals that have guided national and international environmental policy. Having done this we shall then consider whether these ecological goals complement or conflict with two other major goals that are given great emphasis in the global community: the maintenance of international peace, and the economic development of the Third World.

Basic Environmental Values

Those who embrace environmental values are not of one mind in agreeing on what could be called an environmental ethic. The concept of environment is an umbrella term that covers many aspects of man's relationship with his natural surroundings. Which aspects are given priority depends upon one's circumstances and interests. To a backpacker, the key environmental value may be preserving the pristine character of wild areas by designating them as wilderness. The primary environmental concern of a middle class commuter in a smog-choked urban area is likely to be a reduction of noxious air pollutants. To the residents of rural villages in the Third World, the priority may be reducing the prevalence of killing and debilitating diseases, such as those caused by a contaminated water supply. The nomadic herdsman in arid lands is likely to be most concerned about the spread of desert conditions. What follows is a discussion of four of the most prominent environmental values, including (1) controlling pollution, (2) preserving genetic diversity, (3) conserving natural resources, and (4) limiting population growth, and a briefer consideration of several less salient ones.

Controlling Pollution

Man can be a messy creature with his habit of discarding his wastes wherever convenient—into the atmosphere, rivers, and oceans, or onto or beneath the surface of the land. Any such substance generated or displaced by human activities, whether in the form of a gas, liquid, or solid, which is out of place in nature and, thus, interferes with its processes or has other undesirable consequences, can be referred to as a pollutant. Environmentalism in its narrowest sense has focused on this type of degradation of our natural surroundings. The goal of national and international public policy has been to limit if not reduce or eliminate the flow of pollution into the environment, especially that which could have particularly harmful consequences. A further objective is to clean up the contaminants that linger from earlier human activities, a task that in most cases is very expensive if not impossible.

There are several reasons why there has been broad-based support for controlling pollution. One is the unsightliness of pollution.

Landscapes strewn with litter and garbage, air clogged with smog, and rivers murky with chemicals are unpleasing to the eye and nostril and obscure the natural beauty of an area. The Japanese, for example, lament the view of Mount Fuji from Tokyo being lost to smog. In major cities, pollutants blacken and even corrode the surfaces of buildings, thus marring the appearance of both ancient and modern structures and cityscapes in general.

If appearances were the only consideration, however, pollution would be a nuisance and controlling it would warrant little priority. Pollution can have many serious consequences, among which are harmful effects on human health. The link between pollution and health problems is clear, even if it is a slow and difficult process to prove scientifically. For example, there is now evidence that air pollutants such as sulfur oxides, carbon monoxide, and particulate matter cause or exacerbate lung cancer, coronary heart problems, and respiratory diseases, including chronic bronchitis and emphysema. Untreated sewage is a major conveyor of diarrhea, cholera, typhoid fever, and debilitating parasites.[3] Heavy metals contained in industrial effluents have toxic effects on certain organs of the human body. An especially tragic example of heavy metal poisoning occurred in Japan where many of the residents of Minamata developed a serious ailment of the central nervous system that was caused by eating fish contaminated with mercury that had been discharged from a local industrial plant.[4]

Controlling pollution is also of concern because of the impact it can have on other species. Among the most vulnerable are several species of birds, such as the peregrine falcon, the bald eagle, and the osprey, whose reproduction has been hampered by the effects of DDT on the thickness of their egg shells. Abnormally large amounts of phosphorous in bodies of water receiving fertilizers in the runoff from agricultural land stimulate algae growth, which consumes the oxygen essential to the survival of other types of marine life, including fish. As a result of this process, known as eutrophication, lakes and other bodies of water become virtually dead in a biological sense. One of the most celebrated cases of eutrophication is Lake Erie, which has recently begun coming back to life following concerted efforts of the United States and Canada to stem the flow of agricultural pollutants.

The potential impacts of pollution on climate also pose serious problems. Large amounts of particulate matter may deflect solar energy, causing a cooling of the atmosphere that could shorten growing seasons in key agricultural areas. The reverse is also possible: that a buildup of carbon dioxide in the atmosphere will trap heat in the lower atmosphere causing it to gradually warm up. This process, known as the greenhouse effect, could cause the polar icecaps to melt more rapidly, thereby raising the level of ocean waters and flooding coastal cities in addition to altering climatic patterns in ways that could adversely affect agricultural production.

Preserving Genetic Diversity

Homo sapiens is not unique among species in having the capacity to bring about the demise of other species. The human race does, however, stand out from all others in the sheer numbers of species that it can destroy, either by hunting and killing their members purposefully or by altering their habitat in ways that doom them to extinction. Preventing the future extinction of the millions of species of plants and animals that inhabit the planet has been one of the principal objectives of environmental policy of the United Nations Environment Program and several national governments.

As in the case of controlling pollution, there are aesthetic reasons for trying to prevent extinctions. Many of the endangered species of flowers, birds, butterflies, fish, mammals, and other plants and animals are notable for their beauty. Even tiny organisms that are hardly noticeable to the naked eye can be captivating when viewed through a microscope. The whale, on the other hand, has long held a fascination to man because of its size, apparent intelligence, and advanced ability to communicate. Other animals are admired because of their speed or grace. The loss of very many of these species would irreversibly diminish the quality of life for man.

A second argument for preserving genetic diversity is based on one of the fundamental principles of ecological theory: a diverse group of species contributes to the stability of natural systems. The survival prospects for each species are enhanced when there is a variety of types of food available. Likewise, a more diverse group of predators assures that the population of any given species will be kept in line.[5] Certain bird and mammal species perform the role of the miner's canary in being especially vulnerable to environmental changes and, in so doing, offer an early warning of threats to other species and possibly even to man himself.

Genetic diversity is a key resource in the struggle to provide food for the rapidly growing population of the world. Of an estimated 80,000 edible plants, only 3,000 have been used for food and of these only 150 have been cultivated on a large scale.[6] Some of these unused species, if preserved, may prove to be significant sources of food. In this regard, the great increases in world food production that have barely kept pace with mushrooming population growth rates over the past three decades were made possible by the use of wild strains to breed high-yield, "miracle" varieties of wheat, rice, and other crops. Plant geneticists are also constantly on the lookout for strains that are resistent to the diseases that eventually become adapted to commonly cultivated varieties of the basic food crops upon which there is such a heavy reliance.[7] Thus, preserving genetic diversity, either in the wild or in genetic banks, can be essential to long-term human welfare.

Certain species of plants and animals have been of value to man in other ways. Major ingredients for a whole host of drugs are derived from plants. Certain species of animals have proven to be useful in medical research, especially those that are susceptible to the same diseases, such as cancer, that man is. Substances from plants and animals have also been widely used in the manufacture of industrial products such as lubricants, glues, waxes, rubber, insecticides, fibers, and cosmetics.[8]

Are certain species of greater value and, therefore, more worthy of efforts to conserve them? Public interest lobbies have emerged to promote the cause of the whale, dolphin, fur seal, and sea turtle as well as various species of birds and African wildlife. But are these large, better-known species of mammals and birds more important than some of the millions of lesser-known species of insects and plants, including many that have not yet been categorized, any one of which may have a genetic trait that could be the key to breakthroughs in agricultural production or a medical treatment? Until more is known about the potential of these species, it would be premature to suggest that any are expendable.

Slowing down the alarming rate of extinctions will require concerted efforts to preserve the habitats essential to the survival of endangered species. This was one of several messages of the recently adopted World Conservation Strategy. The greatest concentrations of endangered species are found in tropical forests, which in some areas have been shrinking rapidly as a result of population growth and development efforts. Other endangered species depend upon habitats as diverse as wetlands, deciduous forests, grasslands, deserts, high mountains, and tundra. Numerous nation-states have recognized the importance of preserving these habitats by setting them aside as national parks or nature reserves. Several international projects have been undertaken to identify and to protect critical habitats, the most notable of which is UNESCO's designation of Biosphere Reserves that now number 193 in fifty countries.[9]

Conserving Natural Resources

Human beings have always helped themselves to the natural bounty of the earth in their struggles for survival and their efforts to improve their quality of life. Whatever they find to be useful toward these ends is known as natural resources. It should be borne in mind that the inventory of resources is continually undergoing change as uses for previously neglected substances have been discovered. Moreover, what was once a resource may no longer be of much utility as substitutions are made that serve a purpose more effectively or economically. Insuring that the natural resources essential to human survival and a decent standard of living will be available long into the future has been one of the primary objectives of environmentalists. Currently living generations are being exhorted to leave their descendants a planet whose natural resources have not been recklessly plundered.

The challenge to conserve is slightly different depending upon whether a resource can be classified as being renewable. Renewable resources are those that can be used repeatedly without being depleted. Water, agricultural land, forests, and certain wild species of animals, such as fisheries, are among the principal renewables. These resources play a key role in meeting humanity's food needs and, in the case of forests, provide construction materials, a source of energy, and a raw material in numerous industrial processes. Most renewables are susceptible to overuse or misuse, which jeopardizes their future as a resource. Thus, conservation implies careful use of them so that their value to humanity is not diminished. Agricultural land, for example, is conserved when it is cultivated in a way that minimizes the erosion of fertile topsoil and when crops are rotated to maintain the nutrient level of the soil, even though short-term profits may be sacrificed. Conservation of forests implies that they are managed for a sustained yield by avoiding practices such as clear-cutting on steep slopes that results in the erosion of hillsides. Fisheries are conserved when a sufficient number of fish are left to regenerate the stock to previous levels. Serious air or water pollution can also undermine or destroy the regenerative capacity of any of these biological systems.

Nonrenewable resources are a one time endowment in the sense that nature does not replenish them in a useful form during a time span that is relevant to human activity. Among the most important of the nonrenewable resources are fossil fuels, which have been a critical source of energy for transportation, heating, and industrial purposes, and minerals, which are used as raw materials for most of the industrial products that have enhanced the standard of living especially in the developed world.

In the case of these nonrenewables, conservation implies being parsimonious in their use so that the accessible reserves are not rapidly used up, leaving little if any for future generations. This is sometimes referred to as a slowing down of the process of entropy (whereby matter and energy are transformed by human activities from the concentrated form in which they are found in nature and therefore useful, to something that is dispersed, or entropic, and therefore of little or no utility to man).[10] This can be accomplished by cutting down on wasteful consumption of natural resources, such as by reducing the number of gas-guzzling automobiles or by encouraging more adequately insulated buildings. Conservation may also mean the use of more plentiful substitutes, ideally of a renewable nature, for resources that are in short supply. For example, one can use solar energy for home heating rather than electricity generated from dwindling reserves of petroleum. In the case of minerals, recycling of metals may significantly retard the depletion of natural resources.

Limiting Population Growth

Human beings have used their innate intelligence to surmount some of the more immediate natural factors that would otherwise retard their growth in numbers. Among the most notable of these accomplishments are the invention of weapons to keep natural predators at bay, the expansion of the food supply to reduce the prevalence of famine, and the prevention or treatment of diseases that previously shortened human life spans. The resulting growth in human population, which has been especially pronounced over the past several decades, lies at the root of many of the environmental problems that have been discussed thus far. A larger population is likely to deplete nonrenewable resources more rapidly, to overuse renewable resources to the point of diminishing their value, to add to the burdens of pollution on the ecosystem, and to destroy the habitat of other species. For these reasons, bringing population growth under control has been one of the principal ingredients of strategies for coping with a wide range of environmental problems.

Biologist Garrett Hardin has been one of the most outspoken advocates of limiting human population in the interests of preserving the environment. In his controversial theory of "lifeboat ethics," Hardin goes so far as to suggest that international food assistance should be denied those societies whose population has overshot their capacity to provide for their basic food needs. He would withhold food assistance even during periods of famine due to drought or other climatic calamities, which he suggests "serve to prune away the luxuriant growth of the human race."[11] Providing food assistance is counterproductive because it dampens whatever incentives a society has to limit its population. Hardin counsels us not to confuse the future survival of the human species with the survival of every member of the current generation of the human race. As paradoxical as it may seem, trying to keep everybody alive today may undermine the long-term survivability of the species. Hardin summarizes his argument as follows: "I do not see how we can do any good at all unless we make carrying capacity the primary ethical consideration, putting human lives in a subservient position. This conclusion follows not from a contempt of human lives but from a concern for humanity that extends beyond the passing moment of the present."[12]

Some observers criticize Hardin for emphasizing population growth in the Third World as being the greatest threat to the environment. To them population is not so much the problem as what they consider to be the profligate overconsumption of resources in the industrial world, which they believe has had a far greater impact on the environment.

There is, however, widespread recognition throughout most of the world of the importance of limiting population. How it is to be accomplished raises value questions on which there is hard disagreement. Even to many of the most devoted environmentalists, Hardin's "lifeboat ethics" is morally repugnant because of its apparent callousness to less fortunate human beings who individually bear little if any responsibility for the overpopulated conditions in their countries. Family planning is a more palatable strategy for controlling population, especially if it is directed toward preventing pregnancies voluntarily. There is, however, strong resistence among some groups, especially in the United States, to family planning by aborting pregnancies, which with 40 million being performed worldwide each year has become the third most prevalent form of birth control behind voluntary contraception and sterilization.[13] The public acceptance of any strategy of family planning may also diminish sharply when attempts are made to make it compulsory as has been the case in a few countries, India and China being the most notable examples.

Other Environmental Values

Several other less dramatic environmental causes should be briefly noted. Wilderness advocates argue for preserving some areas in their pristine state so that human beings will have an opportunity to observe what the natural world was like before being altered by man. For some, wilderness is a sanctuary from the congestion, bustle, and ugliness of urban life that can be used for relaxation and achieving a sense of spiritual renewal. A related environmental value is preventing the despoiling of places of outstanding natural beauty, such as mountains, canyons, fjords, waterfalls, free-flowing rivers, and coral reefs; or that have unusual qualities, such as geyser fields, peculiar rock formations, or gigantic trees. These latter types of natural wonders, which are typically incorporated into national parks, may be of interest not only to residents, but to tourists from other countries where there is nothing comparable.

Limiting the harmful consequences of extraordinary natural events, such as floods, landslides, hurricanes, earthquakes, volcanic eruptions, and droughts, has also become a goal of national and international agencies. While these occurrences usually cannot be prevented, means are being developed that will offer improved prior warning, which when combined with precautionary measures, such as not building in flood plains, can reduce the destruction that is left in their wake.

There is also a growing consciousness of the need to anticipate the environmental consequences of large-scale human projects. Rerouting rivers, for example, could have major environmental ramifications, some of which are difficult to foresee. A case in point was a Soviet plan to alter the course of Siberian rivers that now flow into the Arctic Ocean to make them flow southward to supply water for irrigating agricultural lands. Irrigation projects that transport water from the lake behind the Aswan High Dam in Egypt have provided an ideal environment for the spread of parasite-infested snails that carry the painful energy-sapping disease schistomsomiasis. Recently, more sensitivity has been shown for the effects that these projects can have on the life-styles of native peoples, such as the Amazonian Indians whose domain is being encroached upon by large-scale land-clearing operations in the tropical forests of Brazil, the Lapp people who live in an area of northern Norway where a major hydroelectric project has been planned, and the Eskimos of the north shore of Alaska whose homeland has been intruded upon by oil development.

The concept, environment, has also been stretched to include the surroundings in which people live or what in international circles is known as human settlements. Since the Vancouver conference in 1976, improving human settlements has become one of the primary missions of the United Nations Environment Program. Among the key values contributing to a minimally adequate quality of life in human settlements are a safe supply of drinking water, sanitation systems, adequate shelter, convenient and affordable transportation, and other civic services, such as police and fire protection. The environment of workplaces is also being scrutinized to prevent exposure to toxic substances, excessive noise, and other safety and health hazards.

This concludes a brief overview of the values that have been guiding international environmental policies. Let us now consider these environmental priorities in relation to some of the other principal goals of the international community.

Peace and the Environment

Peace has traditionally been the preeminent goal of the international community. Article I of the Charter of the United Nations, which was adopted shortly after the sobering experience of World War II, lists the maintenance of international peace and security as the first of the objectives for which the organization was formed.

The concept, peace, can be interpreted in a variety of ways. Defined most narrowly, it is the absence of violence, which in the international community is usually taken in the form of military combat or warfare. Defined most broadly, international peace implies not only the absence of war but also an end to preparations for war on which nearly $660 billion a year is spent globally, a figure that is comparable to the combined income of the poorest half of the world's population.[14] Without general disarmament it is feared that peace can at best be a fragile, temporary state of affairs, which during a crisis could quickly deteriorate into devastating military conflict. When defined in terms of what it is, rather than what it is not, peace is described as harmony in the relations between states and international cooperation in addressing problems of common concern.

What is the relationship between peace and the environment? One school of thought, which now has few proponents, is based on the dubious proposition that war has traditionally been one of nature's ways of keeping the human population under check. Even during eras of widespread and intense warfare, however, only about 2 percent of the deaths taking place could be attributed to such hostilities.[15] Most evidence would lead toward the opposite conclusion that wars and arms buildups generally have a harmful or even disastrous impact on the natural environment and that harmonious relationships between nations facilitate the cooperation necessary to address the environmental problems that cannot be effectively addressed by countries acting on their own. Reserving the direction of cause and effect, it is also evident that environmental problems, in particular scarcities of natural resources, can be a major cause of international tensions leading to warfare. Thus, on the whole, peace and environmental goals would appear to be mutually reinforcing.

As was suggested in the introductory chapter, a major exchange of nuclear weapons could have a catastrophic effect on the environment, rendering much of the planet unfit for human habitation due to radioactive fallout. Limited warfare fought with conventional weapons can also cause significant damage to the natural environment, a case in point being Vietnam, where decades of warfare have ravaged the environment. The usefulness of much of the fertile land of the country was reduced if not destroyed by the saturation bombing of the United States, which pockmarked the countryside with thousands of craters, causing the land to look like a lunar landscape. Brackish water accumulating in the craters provides an ideal environment for disease-bearing mosquitoes. The impact of the exploding bombs disturbed the layer of topsoil leading to an irreversible hardening of the top layers. Long after the war, unexploded munitions continue to pose an ever present threat to farmers plowing their fields. This damage was an inadvertent consequence of a military tactic designed to interdict the flow of men and supplies from North to South Vietnam. Intentional destruction of the environment also became a tactic of warfare when large forested areas were systematically defoliated using herbicides to deny insurgent forces the cover offered by the thick jungles of the region. Toward this same end, huge bulldozers known as "Rome plows" were used to clear large areas of land. The resulting health risks and environmental losses will continue for decades. Efforts were also made to modify the weather by cloud seeding in order to keep roads and trails muddy as a way of hampering the flow of vehicles carrying supplies down the Ho Chi Minh Trail to the south.[16]

Preparation for war can also have adverse environmental impacts. Prior to the Partial Test Ban Treaty of 1963 there was growing concern that atmospheric tests of nuclear explosives were gradually poisoning the environment with radioactive substances up to levels that could be dangerous to human health. Most recent atmospheric tests by China and France have rekindled these fears. Original plans for deploying two hundred of the new MX missiles, each of which would have twenty-three separate potential launch points, drew heavy criticism because of the disruptive impact that they would have on the environment of large areas of Utah and Nevada. Preparations for conventional warfare also have substantial environmental costs, in particular in the consumption of scarce resources, such as petroleum and various strategic materials.

The influence of environmental factors on the prospects for peace or war is perhaps more pronounced. Wars have been fought throughout history over access to resources, an especially destructive example being World War II. Consider the following quotation from Benito Mussolini, the Fascist leader who led Italy into the war.

> We are forty million, squeezed into our narrow but adorable peninsula, with its too many mountains and its soil which cannot nourish so many. There are around Italy countries that have a population smaller than ours and a territory double the size of ours. . . . It is obvious that problems of Italian expansion in the world is a problem of life and death for the Italian race. I say expansion, expansion in every sense: moral, political, economic, demographical.[17]

Resources were also a major consideration in Hitler's ambition to conquer Europe, as the following quotation from one of his speeches in 1936 suggests.

> If we had at our disposal the Urals, with their incalculable wealth of raw materials, and forests of Siberia, and if the unending wheat fields of the Ukraine lay within Germany, our country would swim in plenty. . . . Where is there a region capable of supplying iron of the quality of Ukrainian iron? Where can one find more nickel, more coal, more manganese, more molybdenum? With the addition of this area, we shall be the most self-supporting State, in every respect, including cotton, in the world. Timber we shall have in abundance, iron in limitless quantity,

the greatest manganese ore mines in the world, oil—we shall swim in it.[18]

Japan, a third aggressor in World War II, was also striving for control over the natural resources needed to feed its rapidly growing population and to fuel its industrial development. Japan's capacity to import needed raw materials was undermined by a sharp drop in its exports caused in large part by the collapse of the American silk market in the early 1930s.[19] This unreliability of international trade spurred the Japanese into seeking an empire in East and Southeast Asia.

War over resources continues to be a real possibility today. Tensions in regions such as the Middle East and the southern part of Africa are complicated by international dependence on the petroleum and mineral wealth of these areas and the readiness of outside powers to intervene militarily if the flow of vital resources is thought to be in jeopardy. As world population increases, along with the demand for more resources, the potential for conflict over them also grows.

These circumstances have led to a rethinking of the meaning of security. It can be argued that security is no longer simply a question of having the military might needed to deter or defend against an attack. Security also implies, among other things, reliable access to the food and natural resources needed to sustain the life-style to which a society has been accustomed or to which it aspires.[20] It is in this sense that strategies for preserving nonrenewable resources and conserving renewable ones contribute to the achievement of international security and ultimately peace.

Economic Development and the Environment

During the past fifteen years, the rise in environmental consciousness centered in the developed world has been paralleled by rising demands from Third World nations for increased cooperation and assistance from the international community in achieving their aspirations for more rapid development. These demands have taken the form of calls for a "new international economic order," as outlined in the Charter on the Economic Rights and Duties of States, adopted by the General Assembly in 1974 and, more recently, in the call for "global negotiations" that would lead to its implementation. In contrast to the complementary relationship noted between peace and environmental values, there has always been a certain tension between the goals of environmental preservation and economic development.

In the developed world, there is widespread but rarely expressed concern about the consequences for the natural environment if the 70 percent of the world's population that lives in the Third World were to achieve a modern industrial life style. How long, for example, would remaining reserves of petroleum hold out if 1.5 billion Indians and Chinese were able to move from bicycles to automobiles as a primary means of transportation and to mechanize their agriculture rather than depend on the labor of human beings and draft animals? If they were able to burn large amounts of coal to provide for the desire for rapidly expanding electricity, would this not introduce large amounts of carbon dioxide into the atmosphere hastening its warming due to the greenhouse effect? Were they able to adopt a meat-oriented diet, would the world's food growing capacity not be badly overtaxed? These are but a few examples of numerous environmental problems that are feared as a consequence of the economic development of the Third World.

In contrast, in the Third World, environmental quality has traditionally taken a back seat to addressing the more immediate problems of widespread poverty. Inequities in the distribution of the world's wealth, highlighted by the worldwide spread of the mass media, has given rise to a prevalent belief that it is now the turn of the less developed peoples to pollute and consume resources in the same way that the industrial world has done to achieve its standard of living. As a consequence, the initial reaction by much of the Third World to environmentalism has been disinterest, occasionally punctuated by open hostility generated by concern that the attention of the developed world will be drawn away from the development issue. Moreover, the interest of the North in providing assistance for family planning programs has been widely interpreted as an effort to keep the world's rich minority from being overwhelmed by the sheer numbers of the more rapidly growing population of the less developed regions.

The possible impact of strict environmental standards within the developed countries has also been a cause for concern in less developed countries. If adopted by developed countries, such restrictions could make it more difficult for less developed countries to penetrate their markets, as would be the case, for example, if strict limits were imposed on residues from chemical herbicides and pesticides on agricultural commodities. If environmental restraints are imposed, the price of imported goods might then increase as manufacturers in developed countries pass along the costs incurred in conforming to the regulations. Rules that offer incentives for recycling could also reduce the demand for primary commodities upon which Third World countries rely for export income. Finally, there has been concern that lesser developed countries would be pressured by international institutions such as the World Bank to comply with strict antipollution standards requiring costly investment of their scarce capital and making their countries less attractive to industries seeking a haven from environmental regulations.[21]

As described in chapter 2, the lack of enthusiasm of the Third World for the environmental agenda of the developed world first surfaced prior to the 1972 Stockholm Conference. For some time the conveners of the conference were concerned about reports that many of the Third World countries would boycott or, if they did attend, would obstruct efforts to adopt an international environmental program. Anticipating these possibilities, a preparatory meeting was held in Founex, Switzerland, in 1971 to address the reservations and concerns of the Third World. The meeting contributed to a reconciliation of the perspectives of North and South, which in turn has made it possible for the international community to pursue an active environmental program. This meeting of the minds is reflected in the first article of the Text of Environmental Principles adopted at Stockholm which states that: "Man has the fundamental right to freedom, equality, and adequate conditions of life, in an environment of a quality which permits a life of dignity and well-being. . . . In this respect, policies promoting or perpetuating apartheid, racial segregation, discrimination, colonial and other forms of oppression and foreign domination stand condemned and must be eliminated." The relationship between development and the environment was taken up again in a symposium sponsored by UNEP and UNCTAD, which drew up the Cocoyoc Declaration of 1974. This document strongly criticized the pattern of overdevelopment in the affluent countries and underdevelopment in the impoverished societies and gave more specific meaning to the term eco-development.

As a concession to these concerns, the developed countries agreed to place the headquarters of the United Nations Environment Program in Nairobi, Kenya, the first such UN agency to be located in a Third World country. This decision offered some assurances that the perspectives of the Third World on environmental issues would not be ignored.

In the decade since the Stockholm Conference the tendency to think of development and environmental goals as being mutually exclusive has given way to an awareness of ways that these two priorities are reinforcing. Underlying this change in thinking is the realization that some of the same environmental problems that have been plaguing the industrialized countries are becoming increasingly serious in Third World countries, especially those that have undergone rapid industrial growth. The most visible of these environmental problems is air pollution, especially in major urban concentrations, caused by automobiles and other motorized vehicles and by industrial operations. Third World cities including Mexico City, Cairo, Calcutta, Beijing, Ankara, and São Paulo have some of the highest levels in the world of pollutants. In some cases, foreign firms have located polluting industries in less developed countries because of stricter environmental laws in the developed world that would hamper if not prohibit their operations there. Political leaders in Third World countries have become increasingly aware of both the short- and long-term consequences of pollution and have expressed interest in avoiding some of the mistakes made by the industrialized countries when they underwent rapid development.

Some environmental problems are far more prevalent in the less developed than in the industrial countries. These problems are largely a consequence of a low level of economic development. Such is the case with deforestation, which has been proceeding at an alarming rate in much of the Third World as poor people, unable to afford other forms of energy, strip wooded areas for firewood. Likewise, human sewage is allowed to contaminate streams because there is a lack of resources for constructing treatment facilities. Dangerous pesticides and herbicides are often used in agriculture in less developed countries because there is also a lack of expertise on the dangers associated with using them. Also, not to be overlooked is the tendency for population growth to be the highest among the poorest people where pronatalist (in favor of more children) tendencies are a social adaption to high rates of infant mortality and a lack of social security programs that leaves children as the principal means of support for aged parents. In all of these cases, progress on the environmental front can be expected to go hand in hand with progress in reducing rural and urban poverty.

Growing knowledge of the fragile relationship between humanity and its natural surroundings has led to a reexamination of strategies of economic development and even the meaning of the concept itself. The Western model of development emphasizing rapid industrial growth entailing intensive consumption of natural resources is ill suited to an era in which key resources, in particular those relied upon for energy, are becoming scarce and much more expensive. It is becoming evident to Third World leaders that improvements in living standards will be based more and more on achieving greater efficiency in the use of natural resources. Moreover, with the benefit of hindsight, many leaders in the Third World have expressed a desire to avoid some of the serious types of environmental degradation that accompanied the industrialization of the West, which are either irreversible or very costly to correct. Finally, in an increasing number of Third World countries, China and Mexico being two notable examples, there is a realization that rapid population growth cancels out the potential impact of economic growth on living standards and places heavy strains on the environment. Thus, development, rather than simply being an increase in GNP and industrial output, is being looked upon as a composite of improvements in the quality of life for a smaller population, which because it is accomplished in an environmentally responsible way is sustainable into the future.

Summary

In this chapter we have considered environmental values in terms of what constitutes an appropriate relationship between man and his natural surroundings. Some of the key specific values are (1) the control of pollution, (2) the preservation of genetic diversity, (3) the conservation of both renewable and nonrenewable types of resources, and (4) the reduction of a high rate of world population growth. The strength of the international commitment to these environmental values—which have emerged relatively recently—depends on their compatibility with other widely accepted and long-standing goals of the world community. In this regard, it is encouraging to note that there is in general a complementary relationship between peace and environmental objectives. The compatibility between environmental values and economic objectives has been in more doubt, but over the past decade the dialogue between North and South has done much to achieve a reconciliation between what were originally believed to be competing values.

4 Environmental Policies

Environmental problems pose a variety of serious challenges for international policy makers. Many of these problems and the potential solutions to them are similar in their basic form to the now familiar metaphor, "the tragedy of the commons." In this chapter we shall use the relatively simple and easily understood scenario of the "tragedy" as a way of introducing the more complex problems of the global environment and presenting the principal policy options for addressing them. After doing so, four case studies of actual international problems and policies will be considered: (1) the living resources of the oceans, (2) the extinction of terrestrial species, (3) oil pollution in the oceans, and (4) transboundary acid rain. Each of these case studies begins with background information on the problem, then reviews the specific policies that have been adopted by the international community for addressing it.

The Tragedy of the Commons

Imagine an old English village with a community-owned green. The green serves as a pasture upon which the resident herdsmen of the village are permitted to graze cattle for their own personal profit. Such an arrangement, known as a commons systems, is quite suitable so long as the combined herd of the villagers does not exceed the carrying capacity of the pasture. Recall that carrying capacity is defined as the number of cattle that can be grazed by the pasture without depleting the grasses at a faster rate than they are regenerated. Let us imagine further that the villagers are continually adding to their personal herds. Eventually the carrying capacity will be reached and surpassed. As a consequence of the overgrazing, the cattle then begin showing signs of being undernourished and each head of cattle returns slightly less profit to its owner. Even so, the villagers continue adding cattle to the pasture. Eventually, the grasses are all consumed and the once verdant pasture will no longer support any cattle. The value of the green to all alike has been destroyed. The "tragedy of the commons" has run its course.

How likely would such a turn of events be? What would motivate the villagers to continue adding to their herds even after it became obvious that the pasture was being overgrazed and was well on the way to complete destruction? To answer this question it is necessary to picture yourself in the position of a typical villager. Let us assume that you have a sufficient amount of money to purchase an additional cow. You have the right to graze the cow on the village pasture free of charge, yet all of the profits from the dairy products that are produced will accrue to you personally. Under these circumstances economic incentives for adding to your personal herd are obviously quite strong.

Would your decision be different, however, if you noticed that the pasture was showing signs of becoming overgrazed and that the production of the cattle you already graze on the pasture was declining slightly? Perhaps. But you calculate that the added profits will be far greater than whatever slight reduction there may be in the profits from your other cattle due to their contribution to an overgrazing of the pasture. The total costs of the additional overgrazing may in fact exceed your added profit, but your share of the costs is inconsequential because you share them with all of the other villagers.

If it were only one villager adding a single cow, the consequences for the community would not be especially serious. Unfortunately, the calculation of personal gain applies not only to the first additional cow, but also to a second, a third, and so on. Moreover, the calculations that motivate your own decision to add more and more cattle may also motivate others to do likewise. Thus, under such a commons arrangement, it is only a matter of time before the "ultimate" tragedy takes place.

Averting a Tragedy

Several policy approaches have potential for averting a "tragedy of the commons." These approaches are restraints, restrictions, taxes, enclosures, and public monopoly.

Restraints If the villagers are aware that adding more cattle will ruin the pasture, is it not possible that they will voluntarily limit the size of their herds? If they are not fully aware of the dangers of the impending tragedy, perhaps all that is necessary is to educate them on what is happening and to persuade them of the importance of exercising restraint. The villagers could be given a guideline on the maximum number of cattle each could have without exceeding the carrying capacity of the pasture. Exhortations for voluntary restraint could be couched in terms of responsibility to the community. Social pressures could be exerted to further encourage compliance. Many traditional societies have rituals or taboos which exert such pressure to restrain members from overexploiting their environment.

Restrictions Anticipating that voluntary restraints will be unsuccessful, the villagers may elect to take stronger action in the form of enforced regulations limiting the number of cattle that may be grazed. A maximum could be set for the entire community and villagers allowed to add to their personal herds on a first-come-first-served basis until the quota was reached, after which no additional cattle would be allowed. Alternatively, an individual quota could be set for each villager. These quotas could be set to allow all to have an equal number of cattle or to reflect the varying number of cattle that the villagers were grazing at the time the limits were enacted. Provision could be made for villagers grazing less than their allowed number to sell their unused quota rights to others. Alternatively, all quota rights could be sold to the highest bidder. Villagers who disregarded the limits would be fined by a village authority and their excess cattle would be confiscated.

Taxes Rather than set absolute limits on the numbers of cattle that could be grazed, the villagers may adopt measures that reduce, if not eliminate, the economic incentives for building up large personal herds. A tax on cattle would be one such strategy. The first few cattle of each villager would be taxed at a low rate if at all. Additional cattle would be taxed at progressively higher rates, thus reducing the profit that would accrue to the herdsmen by adding to their herd to the point that they would realize little if any income from additional heads.

Enclosures A fourth strategy for averting the tragedy would be to discard the commons arrangement altogether. The pasture would be divided into sections with one section being assigned to each villager, which would thereafter become his private property. It would be up to the villager to decide how many cattle he would graze on his section, keeping in mind that he would personally bear all of the costs of overgrazing. Such an arrangement incorporates the principle of "intrinsic responsibility." This means that those who benefit from using a natural resource bear the full consequences of its overuse or misuse. Thus, the herdsmen have an added incentive for acting in an environmentally responsible manner. Such a system was adopted in many English villages in the form of enclosure laws.

Public Monopoly The four strategies described thus far all assume that the cattle will be treated as private property. A final potential solution to the "tragedy" would be to transfer ownership of the cattle to a village authority. All decisions about the number of cattle to be grazed would be made by the village authority or by its designated manager. All profits would accrue to the authority, which would distribute them upon the basis of some criteria, perhaps equally to all ("communism") or on the basis of need ("socialism") regardless of individual input. Under such an arrangement both the profit and costs resulting from the addition of cattle beyond the carrying capacity would be distributed broadly throughout the community. Thus, there would be little if any incentive for overgrazing.

Achievement of Policy Goals

An important consideration in the selection of a policy is its potential for achieving the values that are held to be important by those who will be affected by it. The following are some of the goals that the citizens of the old English village might have considered in deciding how to manage the use of their pasture.

Conservation From an environmental point of view the key question to ask in evaluating a policy is how successful it will be in averting a "tragedy" by preventing the grazing of too many cattle on the pasture.

Production The villagers look upon the pasture as a resource that enables them to produce meat and dairy products, which are essential to their livelihood. The greater the production, the higher their income and standard of living. Thus, they are interested in a policy that enables them to achieve the maximum sustainable yield from the pasture.

Equity Discord may arise among the villagers if the policy is not viewed as being equitable. Equity implies justice and fairness, but not necessarily equality in the distribution of income. An equitable policy may take into account differences in the amount that is originally invested or may be based on need determined by family size. Disagreements may arise over which criteria should be used to determine equity.

Freedom As with most of us, the villagers may resent constraints on what they do, especially if these restraints are imposed by a government. At the international level, restraints on the freedom of governments to do as they please may be looked upon as an infringement of their sovereignty.

Implementing the Policies

The policy alternatives that in theory have the greatest potential for realizing the goals of the villagers or the international community may, however, be difficult if not impossible to carry out. Let us briefly consider some of the difficulties that may be encountered in implementing the various strategies we have discussed for averting a tragedy of the commons.

Agreement The first step in pursuing many policies is reaching a common understanding on what is expected from those who will carry it out. In a small village, an informal understanding among its residents may suffice. In larger, less personal communities, the agreements usually are expressed in the form of laws or international treaties or some other formal understanding. Reaching an agreement usually entails negotiations and compromises, which can drag out for extended periods. In some cases, conflicts of interest and mutual hostility, fear, or distrust may thwart efforts to reach an agreement on how to conserve a resource.

Institutions A policy can be considerably more difficult to implement if permanent institutions, such as a village authority or an international organization, must be established and maintained. Some actors may be reluctant to delegate responsibilities and decision-making powers to such an authority fearing that it may take actions contrary to their interests. Even if there is a consensus on the need for such a body, disagreements over the scope of its jurisdiction, how its decisions will be made, how it will be funded, and how disputes will be settled may be difficult to resolve. Continuing conflicts may hinder the future functioning of an institution and divert it from its primary mission.

Free Riders Communities that lack the authority to make decisions that are binding on all members often have their policies undermined by "free riders." These uncooperative members benefit from the sacrifices of others while refusing to agree to limitations on their own behavior. In the case of the English village, for example, a herdsman may continue adding cattle to his private herd which eat the grass that is made available by the willingness of others to accept limits on the size of their herds. Seeing that their sacrifice is futile and incensed that the free rider is taking advantage of them, these socially responsible members may reconsider their commitment to conserving the resource and join in the race to exploit what remains of it before its final destruction. The free rider problem is a persistent one among nations that, being sovereign and therefore subject to no higher authority, cannot be compelled to accept international obligations. It is sometimes possible, however, to put pressure on free riders to conform to community rules by collective action. In the village context, the threat of being socially ostracized may be sufficient; at the international level, warnings of sanctions, such as a trade embargo, may succeed in bringing uncooperative states into line.

Enforcement The success of policies may hinge on whether rules can be enforced on those who are bound by them. The principal obstacles to effective rule enforcement are, on the one hand, an inability to detect violations and violators and, on the other hand, the lack of authority or willingness to apply sanctions strong enough to deter violators. The English villagers may rely on their law enforcement officer to bring violators to justice. In the case of international policies, there is usually no higher authority with analogous police power. Responsibility is placed on the governments of states to police their own behavior and that of their citizens. Generally, compliance with rules under this system of self-enforcement is better than might be anticipated because most governments desire to maintain the credibility of their commitments.

Which Strategy to Choose?

Each of the five policy approaches for averting a tragedy—restraints, restrictions, taxes, enclosures, and public monopoly—has both advantages and drawbacks in regard to fulfillment of goals and ease of implementation. None of the five stands out as the best or worst alternative.

Voluntary restraints are popular with policy makers because they can be readily implemented. There is no need to negotiate agreements, to establish institutions, or to enforce rules. The strategy also ranks high in preserving the freedom of action of individuals or the sovereignty of nations. The primary drawback of the voluntary approach, however, is its vulnerability to the free rider who continues to deplete the resource, thus reducing or cancelling out the positive results of the restraint of others. Therefore, volunteerism is not very effective as a conservation measure. It also does not result in a very equitable outcome if free riders, lacking scruples, enrich themselves at the expense of those who make sacrifices for the good of the community.

Taxes can be an effective instrument for managing a natural resource if they are set at the proper level, something that may require a considerable amount of fine tuning based on previous experience. If too low, they will not be a very strong incentive for conservation; if too high, they will stifle productive use of the resource. The equitability of taxes can be challenged on grounds that the rich will be better able to afford them and therefore benefit from use of the resource. In response, graduated tax rates or redistribution of the tax revenues could be adopted to enhance the fairness of the policy, at least from the perspective of the poor. The taxing approach permits more freedom than do restrictions because the resource continues to be available to those who are willing to pay for the privilege. Instituting and enforcing a tax system will in most cases be a less complicated task than imposing restrictions, which usually must be more detailed. Free riders may be a significant problem if access to the resource cannot be controlled.

Enclosing sections of a resource to be managed by individual members of the community for their own gain can be a clever way of making conservation a matter of direct self-interest. There is the danger, however, that the owner will seek to maximize short-term profits that can be invested in other enterprises at the expense of preserving the resource for a long-term, sustained yield. Because private enclosures would reduce, if not eliminate, the role of the larger community in the management of the resource, implementation is a relatively simple matter. Individuals must, however, bear the expense of securing their section from encroaching free riders. Production may suffer if the new owners lack the interest or means to use their section fully, unless they are willing to rent it to others who will make better use of it. Heavy users of the resource will lose much of the freedom of action that they enjoyed under the commons system, but the result may be a broader distribution of income, and therefore greater equity.

In theory, public enterprises have greater potential for managing a resource for both conservation and productivity. Such a system also makes it possible to distribute the proceeds in equitable ways, but at the price of a drastic reduction in individual freedom and a dampening of incentives for productivity and efficiency. Moreover, major difficulties may be encountered in implementing a public enterprise in view of the elaborate agreements and institutions that would be needed. It is perhaps significant that the seabed mining enterprise agreed upon at the recently concluded United Nations Law of the Sea Conference will be the international community's first experiment with a public enterprise.

The suitability of these five policy approaches also depends in large part on the specific context in which an environmental tragedy is taking place, as will become apparent from the case studies that follow.

Case Study 1: Living Resources of the Ocean

Background

The oceans are the home of a great variety of plant and animal species that form a complex food chain. At the bottom of the chain are microscopic phytoplankton; at the upper end are the bigger, predatory species of fish and marine mammals, including the blue whale, which, measuring 100 feet in length and 200 tons in weight, is the largest animal species ever to have inhabited the planet. Most of the world's fisheries (a term that refers to concentrations of fish or marine mammals that are suitable for commercial harvesting) are located within two hundred miles of continents, where the waters are rich in nutrients that have been washed or blown off the land. Marine life is found in its greatest abundance where there is also an upwelling of nutrients from the ocean depths, as in the case of the cold Humbolt Current, which flows along the west coast of South America.[1] These characteristics explain why only about 10 percent of the area covered by the oceans has the potential for being even moderately productive as fisheries and only .1 percent has the capacity to be highly productive.[2]

Only a small portion of the more than 14,000 species of fish found in the oceans are harvested commercially, either for human consumption or for animal fishmeal. This combination of direct and indirect consumption of fish accounts for 15 percent of human intake of animal protein.[3] Each species has a unique combination of traits that has implications for how it is harvested and the type of management that is needed. Lobsters and oysters, for example, are known as localized stocks that are either sedentary or move only short distances and, thus, live their lives off the coast of a single state. By contrast, highly migratory stocks, such as skipjack tuna and some species of whales, range over large areas of the oceans, crossing through the coastal jurisdictions of several states and the high seas as well. A few anadromous species, such as salmon, live most of their lives in the oceans, sometimes quite far out from shore, but return to freshwater rivers to spawn, where their movement is increasingly being obstructed by such man-made impediments such as dams. A greater variety of species populate the warm tropical waters, but because they are intermixed to a greater degree, harvesting of them is more difficult and less efficient than in temperate regions where prodigous numbers of fewer species are found.

Although man has been fishing for food since prehistoric times, it is only in the twentieth century, and especially since World War II, that fishing has become a major industry. The world catch of fish more than tripled between 1950 and 1970, climbing from 21 million tons to 70 million tons, most of which comes from the oceans.[4] Since 1970 the overall catch has leveled off despite greater investments in fishing operations, a telltale sign of the depletion of ocean fisheries, which occurs when too few fish are left to regenerate the stock to its previous level. The first signs of depletion of certain heavily fished stocks were noted in the nineteenth century off the coasts of northern Europe. Among the species for which there have been dramatic drops in the catch are cod, halibut, herring, California sardine, and anchovy. Whales have also been particularly susceptible to overharvesting because, being mammals, they have few offspring. The bigger species of whales have been the first to decline rapidly in numbers. Current estimates are that only 10,000 to 13,000 of the enormous blue whale remain, which is thought to be only 6 percent of its original (virgin) stock, a figure that brings them dangerously close to reproductive collapse and extinction. The humpback whale has been similarly depleted. Overall, it is believed that the world's whale population is only about one-half of its virgin number.[5]

Technological advancements in the fishing industry are among the principal causes of the relatively rapid depletion of many of these fisheries. The greater strength of new synthetic fibers has made it possible to use larger nets, which are hauled up by powerful modern machinery. The location of concentrations of fish has been facilitated by the use of helicopters, sonar, radar, and other electronic gear. But perhaps more important has been the introduction of floating fish factories ("mother ships"). These vessels have made it possible for nations such as Japan and the Soviet Union to assemble large fishing fleets that can stay away from their home ports for months and even years while intensively harvesting the world's most productive fisheries in distant locations. After depleting a fishery, something that can occur in a few seasons, distant water fleets have the flexibility of moving on to another promising area.

The capacity of the international fishing industry to harvest the living resources of the oceans has for some time now exceeded the supply of desired species. Nevertheless, intensive fishing continues as the operators of big expensive fleets attempt to pay off their large investments. The very nature of the ocean's living resources also counteracts whatever inclinations one or another fishing fleet may have for conservation: the fish they pass up today are more likely to be caught by somebody else tomorrow than to be left to produce new generations for one's own nets at a later date.

Policies on Living Resources

Most of the living resources of the oceans are found in areas traditionally regarded to be the "high seas" and are subject to the doctrine of the "freedom of the seas" that was developed in a treatise entitled *Mare Liberum* written in 1609 by a Dutch jurist named Hugo Grotius. Under this doctrine, which was to become a basic tenet of international ocean law for more than three centuries, all states are entitled to use the oceans as they please, including the unimpeded right to harvest the living resources found in them. The only exception was a narrow band of territorial waters, generally agreed to be three miles in width, over which the coastal nation-state had exclusive jurisdiction. Beyond this three-mile zone, the living resources could not be claimed by anybody until harvested, at which time they became the private property of whoever caught them. Thus, under the freedom of the seas, the resources of the ocean became a textbook example of a commons arrangement.

The freedom of the seas doctrine was all that was needed when the stocks of fish were plentiful relative to the catch of a largely primitive fishing industry. A growing awareness that choice fisheries were being overharvested prompted two types of corrective responses. One was the establishment of international bodies known as fishery commissions, now numbering twenty, to regulate the catch of one or more species in a certain region of the oceans. These commissions are voluntary associations of states having a common interest in conserving a fishery that they jointly harvest. Their membership ranges from as few as two, to as many as thirty.

Each of the commissions has a special set of strategies to prevent overfishing. Some set "maximum sustainable yields" based on scientific research that they or other organizations conduct. Then, any allowable catch is divided among the member-states on the basis of such criteria as geographical proximity to the fishing grounds, historical share of the catch, or any other considerations deemed to be appropriate. Other commissions try to restrict the catch by declaring closed seasons, in some cases when the combined catch reaches a previously established limit. Several of the commissions regulate the type of fishing gear that can be used; for example, requiring that the mesh of nets be large enough to allow the smaller, younger specimens to escape and therefore sustain the species. The strength of enforcement measures also varies. Some rely exclusively on self-enforcement by member-states. Others permit spot checks by foreign or international observers if there is reason to believe that violations have occurred. A problem common to all of these voluntary associations is the lack of any legal authority to compel all states that fish a region to join the commission and submit to its rules.

Since it was formed in 1946, the International Whaling Commission (IWC) has attracted perhaps the most attention of any of these international bodies because of the widespread popularity of the whale. Its membership has grown to thirty-two states, only a few of which—led by Japan and the Soviet Union—continue commercial whaling. The IWC's first conservation strategy, which was adopted in 1947, set a limit on the total number of whales that could be caught on a first-come first-served basis during the next season. It was a significant step because the new limit was only about one-half the size of the catch of the 1937–38 season. Over the next twenty-five years, this general limit was cut back gradually to only one-fifth of the original 1947 level. In 1972 a new management scheme was introduced that set separate limits for each species of whale based on calculations of the maximum sustainable yield of its stock. These limits have also been revised downward annually. A moratorium was also placed on species whose stocks had fallen below a minimally stable level, which initially included the blue, gray, humpback, right, and bowhead whales. In 1979, the IWC banned all whaling from factory ships (except for the plentiful mink whale), a decision that surprised many observers, and declared most of the Indian Ocean a whale sanctuary because of its importance as a calving area. At the 1982 meetings of the IWC a ban on all commercial whaling after 1985 was adopted. Japan and other whaling nations are threatening to defy it on grounds that it is unnecessary and would result in the loss of a substantial number of jobs in an industry that has existed for a thousand years.[6]

The second broad corrective response to the problem of overfishing has involved coastal states concerned that foreign, distant-water fleets were depleting the fishing grounds used for generations by their traditional, shore-based fishermen. Beginning in the late 1940s, a succession of states have unilaterally declared exclusive economic jurisdiction over the marine resources as far out as two hundred miles from their coasts. International conflicts have arisen when coastal states have attempted to exclude fishing vessels from states that did not recognize these claims. The most notable examples have been the "tuna war" involving American vessels fishing off the coasts of Ecuador and Peru and the "cod war" involving the question of British rights to fish in waters claimed by Iceland.

The issue of coastal zones was one of many addressed by the Third United Nations Law of the Sea Conference, which was convened in 1974. The first two conferences in the series, held in 1958 and 1960, had failed to reach a consensus on how wide such zones should be. Recognizing the futility of trying to roll back the unilateral claims of more than one hundred countries to extended territorial waters and exclusive fishing zones, the negotiators agreed upon twelve miles as the breadth of territorial waters over which coastal states exercise jurisdiction. Beyond that there is an Exclusive Economic Zone (EEZ), extending out to a distance of two hundred miles, over which the coastal state has first claim to living and nonliving resources of both the waters and the seabed. If a coastal state is unable to make complete use of the living resources of its EEZ, there is the expectation that other countries, especially the landlocked ones in its region, will be permitted to harvest what remains of the maximum sustainable yield. Thus, the fate of most of the productive fisheries passes to individual nations, some of which are well prepared to manage them, others of which are not and must rely on outside assistance from organizations such as the Food and Agriculture Organization.

International management of the type offered by the fishery commissions is still needed to conserve highly migratory species, which move from one EEZ to another and sometimes into the high seas. The new ocean law is not very specific on this point, suggesting only that the states that fish these species should cooperate in the interests of conserving them. The recently signed Law of the Sea treaty recognizes the right of the coastal states to manage anadromous species that spawn in their rivers even though they spend part of their life cycle in the high seas beyond that nation's EEZ. Without such management rights, coastal states would be reluctant to bear the cost of keeping channels open for spawning fish.

Case Study 2: The Extinction of Terrestrial Species

Background

In the six billion years of its existence, our planet has been home to an estimated 500 million species of plants and animals. Of these, less than 10 million exist today. The remaining 98 percent have become extinct, mostly through the process of natural selection, which was first identified by Charles Darwin a little more than a century ago. At the same time, with changes in the natural environment, species emerge that have adapted to the new conditions more successfully than previously existing species. Under the "normal" conditions that have prevailed for eons, the number of species has grown gradually. In more recent times, however, one species, Homo sapiens, has accelerated the rate of extinctions both by hunting and killing other species for food, profit, or sport and by inadvertently undermining their survivability by destroying or altering the habitats upon which they depend. As a result, extinctions are now outpacing the evolution of new species, thus reducing the biological diversity of the planet. Ecologists warn that we are now in the midst of what could be called an "extinction spasm." During the last quarter of the twentieth century as many as one million species may be lost, an average of one hundred extinctions a day.[7] In this case study we shall concentrate on terrestrial species, especially wildlife, that reside on the territories of nation-states.

Evidence from fossil remains that have been unearthed reveals that the assault of Homo sapiens on the other species began in prehistoric times. During the Pleistocene era, which ended approximately ten thousand years ago, the planet was home to a remarkable assortment of animals, including species of giant mammals and birds. Improvements in human hunting skills toward the end of the Pleistocene coincided with the disappearance of as many as two-thirds of the species of large mammals. Also, the migration of human beings to the western hemisphere is believed to have led to the killing of much of the wildlife that had not developed an instinctive fear of human beings.[9]

That man can bring about the rapid demise of a plentiful species is well illustrated by the cases of the passenger pigeon and plains bison, both of which were found in especially large numbers in North America until the latter part of the nineteenth century. Possibly the most abundant bird ever to exist, the passenger pigeon traveled in huge flocks, estimated to contain as many as two billion birds, which would darken the sky over the eastern United States for days as they passed over. But during the mid–1800s, the number of passenger pigeons declined precipitously. This was both a consequence of the clearing of the great oak and beech forests in which they nested and the large-scale harvesting of the birds by increasingly effective techniques for capturing and killing them to supply a burgeoning market. By the Civil War, all large flocks had disappeared from the East Coast and by the 1880s they were in sharp decline elsewhere. No passenger pigeons have been sighted in the wild since the turn of the century, and in 1914 the last specimen in captivity died in a Cincinnati zoo.[9]

The great plains of North America were once home to an estimated 60 million bison, which was perhaps the most populous large mammal ever to exist.[10] Native American Indians made use of the bison, especially after they acquired horses from the Spanish, but did not kill enough of them to deplete their numbers significantly. As railroads were built across the continent, however, professional hunters moved into the plains and slaughtered the large shaggy animals in mass, primarily for their tongues, which were a delicacy of the day, and hides, which had a variety of uses. What remained of the carcuses was left on the plains to rot. As late as 1870 herds as large as ten to twenty miles in diameter were seen in Arkansas. In 1883 the last significant herd was harvested and by the turn of the century all that was left was a small herd of five hundred, which was nurtured under legal protection to avert complete extinction.[11]

In modern times growing numbers of terrestrial species have become the victims of human activities. Various species of elephants, rhinoceros, tigers, cheetas, linx, chimpanzees, jaguars, falcons, cranes, parrots, and eagles are among the better known creatures whose survival is seriously threatened. It is feared that many thousands of less noticeable species of insects and plants, especially those found in tropical rain forests, will be lost before they have even been identified and categorized. For most species, the greatest threat to survival is the encroachment of rapidly growing human populations on their habitats. This occurs as agricultural and urban areas spread out, as forests are cleared, as swamps are drained, as rivers are dammed, and as chemicals are introduced in the form of pesticides or herbicides. Many exotic species have been harvested to the point of extinction by hunters who supply a thriving international market for wild animals and plants. Some are taken alive to be sold as pets or specimens for zoos; others are killed either for sport or for the parts of their bodies that have economic value such as meat, skins, feathers, shells, or horns.

The animal and plant trade was particularly active during the 1970s, especially in the case of exports from Third World countries going to wealthy customers in the richer developed countries. Because of their limited availability, endangered species can often command a high price in the marketplace, which in turn increases the economic incentive for depleting the dwindling stock even further. Such has been the fate of the rhinoceros, which may be the next big animal to become extinct. Rhinos have been ruthlessly hunted for their horns, which are prized as handles for the ornate daggers worn by men in North Yemen to display their image of masculinity. In China, the horns are ground into a powder incorrectly believed to have the qualities of an aphrodisiac. As a result, 90 percent of the rhinos of East Africa have been killed off and the price of rhino horn has risen to more than $2,000 a pound in some markets.[12] Similarly, elephants are being slaughtered to the verge of extinction for the ivory of their tusks, which can bring $13,000 for a medium-size specimen. Ironically, the demand for ivory has been stimulated even further by the prospect of extinction, a development speculators believe will cause the price of ivory to soar.[13] Live specimens of certain parakeets will bring as much as $12,000 a pair from bird collectors in the United States and Europe.[14]

Thus, the problem of species extinction has in large part become an economic one. Some species are being hunted ruthlessly for the price they will bring on international markets, and the habitats of many more are being destroyed because other uses of land bring a greater immediate economic return than does the continued existence of an endangered species.

Policies on Endangered Species

The international community has gradually come to accept the principle that species are a "common heritage of humanity" even if the only remaining specimens are found exclusively within the political boundaries of a single state.[15] Otherwise, such species would be looked upon as an exclusive possession of states whose governments could dispose of them as they saw fit. The common heritage principle recognizes the interests of the broader community in the preservation of species, which may have unique genetic resources that will prove to be of value to any or all members. Moreover, it puts the government of the country in which the species is found in the position of being the guardian of an international public trust. This doctrine has been given legal expression in UNESCO's 1973 Convention for the Protection of the World Cultural and Natural Heritage, which established a World Heritage Trust to preserve human artifacts and natural areas of "universal value," including the "habitats of certain species of outstanding value." To help protect the biological and genetic diversity of the planet, UNESCO has also undertaken a major project to identify a representative group of biotic communities to be included in a global network of ecological protectorates known as Biosphere Reserves. By late 1981, 210 such areas had been designated in fifty-five countries.[16]

Being thrust into the role of trustee of endangered species has been a heavy burden for some of the less developed countries. A case in point is Tanzania, whose national parks and wildlife preserves, including the Selous Game Preserve, the largest in the world, cover more than one-tenth of the land area of the country. A Tanzanian delegate to a recent conference questioned whether it was right for the world to say to his country that it must guarantee the survival of these parks even if Tanzanians must do without adequate food, clothing, health, or education. He suggested, further, that if the international community is so strongly interested in preventing the extinction of the species of his country, then it should be willing to share in the economic cost of preservation.[17] The World Heritage Trust has very limited funds available for this purpose as do those IGOs, such as the World Wildlife Fund, which focus on wildlife conservation.

The threat to endangered species posed by a flourishing international trade in them received considerable international attention and led to the adoption in 1973 of the Convention on Trade in Endangered Species of Wild Fauna and Flora (CITES), which was originally drafted by the International Union for the Conservation of Nature and Natural Resources. By 1981 the treaty had been ratified by sixty-seven states, including the United States. These states have committed themselves to prohibit all commercial trade in the 449 species now designated as being "most endangered" and to restrict trade in two hundred other species classified as being "less endangered" in accordance with a permit system, which requires exporting countries to demonstrate that the species is not being overharvested.[18] Every two years the parties to CITES convene to update the lists of endangered species. At the 1981 meeting, for example, which was held in New Delhi, three species of great whales—the fin, sei, and sperm—were added to the list of most endangered species.

CITES sets forth obligations for both exporting and importing states. Exporting states are expected to clamp down on the suppliers of animals for export by establishing and strictly enforcing laws prohibiting the killing and sale of the internationally designated endangered species. Fearing a reduction of export revenues badly needed in their fragile economies, the governments of such key exporting states as Mexico and Thailand have been reluctant to make such a commitment. Some of the ratifiers lack the means to enforce their game laws effectively against professional poachers. An especially unfortunate slaughter of hippopotamuses and elephants by idle soldiers took place during the breakdown of public order in Uganda following Tanzania's invasion to oust the oppressive and cruel regime of Idi Amin.

Animal importing states that become parties to CITES are obligated to prohibit the passage through their ports of specimens of endangered species or of their products, unless they are properly documented. It is presumed that the economic incentives for slaughtering or capturing rare species will decline if the market for them can be denied or at least reduced. The success of CITES has been limited by the reluctance of some key importing nations, such as Belgium, Austria, the Netherlands, China, Singapore, and Yemen, to become parties to the treaty, and by the inability of officials in states that do prohibit such trade to distinguish endangered species from similar ones that are more plentiful.[19]

While CITES is the most significant international agreement designed to prevent extinction of territorial wildlife, there are several other treaties that are noteworthy. Birds and other migratory species that cross international frontiers have been a concern of conservationists since the nineteenth century. A series of international treaties, of which the 1979 Treaty on the Conservation of Migratory Species of Wild Animals is the most recent, charges states that are a part of the range of the species with the responsibility for preserving habitats, such as wetlands, that are critical to them. A 1973 treaty prohibits the hunting, killing, or capturing of polar bears except for a few narrowly defined purposes, such as scientific research and conservation, and places a responsibility on individual states to preserve their habitat in the Arctic regions. Five years earlier, members of the Organization for African Unity drew up the African Convention on the Conservation of Nature and Natural Resources to protect wildlife and plant species with which the continent is so richly endowed.[20]

Case Study 3: Oil Pollution in the Oceans

Background

The oceans have traditionally been the ultimate dump for many of humanity's noxious wastes. Some pollutants are pumped into them directly from ships or are discharged from coastal facilities. Others are carried into the oceans by rivers or fall from the atmosphere with precipitation. Among the major types of pollutants that have been released into the oceans are sewage from municipal systems; the herbicides, pesticides, and fertilizers that run off agricultural land; a great variety of industrial wastes such as mercury, lead, and other heavy metals; radioactive materials from nuclear power plants and weapons production; and petroleum and other oily substances, upon which this case study will focus. Until relatively recently, it was thought that the oceans were vast enough to disperse and neutralize these substances without significant harm to the environment. Such complacency is no longer warranted in view of growing evidence of significant damage to the marine environment from pollutants, especially in certain coastal areas.

Oil pollution has become a serious problem with the dramatic increase in the world consumption of petroleum occurring since World War II. Simultaneously, there has been an even more rapid rate of growth in the international trade in oil, most of which is transported over the seas. However, shipping accounts for only slightly more than one-third of the petroleum hydrocarbons entering the oceans. Over half of that pollution comes from land sources, such as industries and refineries. Approximately 10 percent seeps up naturally from the ocean floor, and although offshore drilling accounts for only about 1 percent of the oil pollution, spectacular blowouts such as the ones that took place in the Santa Barbara Channel in 1969 and in the Gulf of Mexico in 1979—the latter being the largest oil spill in history—can have serious impacts on the local environment.[21]

Most of the pollution attributable to shipping results from the operation of the world's gigantic fleet of oil tankers. Relatively small amounts come from other vessels that use petroleum as a power source. Just after World War II, the largest tankers weighed as much as 25,000 tons. The closing of the Suez Canal to shipping in 1956 brought on the era of the supertankers, ships of 200,000 deadweight tons (load capacity), which could take advantage of the economies of scale in making the long trip from the Persian Gulf around the Cape of Good Hope to Western Europe or through the Straits of Malacca to Japan. Over the years larger and larger tankers have been brought into service, the biggest thus far being the 550,000 ton ship *Batillus*, which is as wide as a football field, is a quarter mile in length, and has a draft (depth) comparable to the height of a six-story building (although with the declines in the price and movement of oil in the 1980s, many of these supertankers were left idle).[22]

Over the past fifteen years, various types of accidents involving tankers have resulted in numerous oil spills. Because of their immense size, supertankers are difficult to manuever and are prone to mechanical and structural failures. Collisions become more frequent as waterways are increasingly congested, and groundings are not uncommon where there are treacherous geological features and severe storms. And volatile gasses remaining in emptied tanks have caused a number of spectacular explosions. The first major tanker accident occurred in 1967 when the *Torrey Canyon*, the third largest tanker of its day, ran aground off the Cornwall Coast of England spilling 29 million gallons of oil. The biggest shipwreck in history involved the *Amoco Cadiz*, which spilled 55 million gallons just off the coast of Brittany in 1978 leading to $2 billion in damage claims. Perhaps the most notable tanker accident off the coast of North America took place in 1976 when the *Argo Merchant*, a small tanker by comparison, ran aground and broke up off New England, spilling 7.7 million gallons near the Georges Bank, one of the world's most productive fisheries.[23]

As dramatic as these shipwrecks have been, accidents account for only 5 to 10 percent of the ship-related oil pollution.[24] Much greater quantities of oily substances are discharged intentionally from vessels as part of their routine operations. After the cargo of oil is removed, some of the empty tanks are filled with sea water, which serves as ballast to keep the vessel stable on its return trip. Prior to reloading, the ballast water is pumped out and returned to the ocean mixed with the leftover oil that coated the insides of the tanks. It has also been a common practice on return trips to wash some of the empty tanks to remove sludge and to dispose of the oily residues by discharging them directly into the oceans ("tar" one often encounters on beaches is actually crude oil pumped overboard in this way).[25]

The seriousness of the environmental damage resulting from oil pollution has been a matter of dispute among scientists. It is apparent that much depends upon the type of oil that is spilled and the local conditions. Under favorable circumstances, much of the oil will evaporate rather quickly or be dispersed by wind and wave action. Tar lumps or balls may persist anywhere from a few months to a year before being completely degraded by microorganisms that feed on them. In the Arctic and Antarctic, spilled oil may remain for decades because of the slower rate of evaporation in cold temperatures. In these areas it is possible that the dark spilled oil will absorb more heat than the surrounding white ice and snow cover, causing an uneven rate of melting that could upset the fragile balance of Arctic ecosystems. In most cases, a single massive spill will be more damaging to the environment than oil entering the environment in a persistent, more dispersed manner.

The most obvious casualties of oil spills are waterfowl. In the North Sea and North Atlantic regions it is estimated that between 150,000 and 450,000 birds are killed each year by chronic oil pollution, and that as a consequence a number of species including the puffin, razorbill, and quillemot are threatened with extinction.[26] Since most types of oil float, less damage is done to fisheries, except when fish become contaminated with toxic substances that render them unacceptable for human consumption. Oil washing into coastal marshes can contaminate and kill marine life such as clams, mussels, oysters, and other organisms that live in the sediments and are an important part of the food chain. Scientists are concerned that we still may be unaware of several serious, long-term consequences of oil pollution for marine life.

Oil slicks that wash on beaches can have a serious economic impact on coastal residents who are the unfortunate victims of circumstances that may include incompetence, carelessness, and irresponsibility on the part of tanker operators. Local fishermen may lose their livelihood for an extended period, if not permanently, and news of dirty and oily beaches could cost a coastal tourist industry millions of dollars in lost revenues. Cleanup of slicks can be a major expense that severely taxes the resources of local governments and, moreover, the detergents used in the process may also be damaging to the environment. For these reasons, coastal states near major sea routes have a strong interest in reducing vessel-source oil pollution.

International Policy on Oil Pollution

Historically, the discharging or dumping of polluting substances into the oceans was permitted under the doctrine of the freedom of the seas, which was discussed in the first case study. As recently as thirty years ago, there were virtually no international laws protecting the oceans from this sort of pollution. But as the magnitude of polluting activities increased together with scientific knowledge of their harmful consequences, it became apparent that this laissez-faire policy on ocean pollution needed to give way to regulations that would prevent serious damage of the marine environment.

In the case of oil pollution, most of the attention of international policy makers has been focused on pollution coming from vessels, either as intentional discharges or accidental spills. The center of rule-making activity has been the International Maritime Organization (IMO), formerly called the Inter-Governmental Maritime Consultative Organization, whose convention was agreed upon in 1948. IMO did not, however, come into being as an international institution for another decade until the convention had been ratified by the necessary twenty-one states.[27] As a Specialized Agency of the United Nations, whose membership is now well in excess of one hundred, IMO's domain is commercial shipping. While not one of IMO's original missions, controlling vessel-source pollution has become one of its most important concerns, especially after the *Torrey Canyon* disaster of 1967.

Efforts over two decades to establish an international policy on vessel-source pollution culminated in the International Convention for the Prevention of Pollution from Ships of 1973, otherwise known as MARPOL 73. When ratified by the necessary number of nations, this document will take the place of an earlier IMO treaty on the prevention of pollution of the sea by oil, which was adopted in 1954 and amended in 1962. The new treaty designates "special arenas" that are considered to be particularly vulnerable to oil pollution, and in which all oil discharges from ships would be prohibited. These special areas include the Mediterranean Sea, the Baltic Sea, Red Sea, Black Sea, Persian Gulf, and Gulf of Oman. Outside these areas, limitations have been placed upon the amount of oily substances that may be discharged.

MARPOL 73 also contains regulations on procedures and equipment that are designed to reduce oil pollution. It requires that "load-on-top" procedures be used to reduce oily discharges in removing ballast waters. New ships over 70,000 dead weight tons are to be constructed with separated ballast and oil tanks and standards pertaining to construction, tank size, and stability are to be imposed on all tankers. Shipowners argue that these new regulations will add substantially to the purchase price of the tankers and detract from the operating efficiency of tanker fleets.

Several other treaties address problems related to accidental spills. A long-standing convention, revised in 1960 and 1972, respectively, known as the International Regulations on the Prevention of Collisions at Sea ("rules of the road"), regulates the movement of vessels just as traffic laws regulate the movement of automobiles. The *Torrey Canyon* disaster prompted a 1969 treaty that places strict liability for pollution damage on the owner of the ship transporting the oil. This provision clears up questions of legal responsibility that had delayed the cleanup from previous accidents. The treaty further requires that shipowners carry insurance or other acceptable guarantees to cover damage claims. A 1979 convention establishes a fund to compensate victims of pollution beyond the limits of liability that were specified in the 1969 treaty on liability.

A 1974 Convention on Safety of Life at Sea (SOLAS) and a protocol added in 1978 prescribe a comprehensive set of design measures focusing on vessel safety, which could potentially reduce the likelihood of accidental pollution. Particular attention has been given to steering, navigation, and the dangers of explosion. In 1978, IMO adopted a convention on training that specifies minimum standards for crews that are substantially tighter than the current standards of many countries.

Over the past decade even stronger measures to reduce pollution have been adopted by regional groupings of states that border a common body of water, for example, the North Atlantic, the North Sea, and the Baltic Sea. The United Nations Environment Program has played the role of facilitator for regional groups interested in developing strategies for cleaning up the bodies of water they share. UNEP's most notable success thus far has been with the Mediterranean states, which have adopted a number of protocols, known as the Barcelona Conventions, that regulate various types of pollutants. Regional plans of action have been adopted for the Caribbean Sea and the Persian Gulf, through which much of the world's oil trade passes. A notable feature of these regional plans are the provisions calling upon the coastal states to reduce pollution from land sources, something that goes beyond the scope of any of IMO's treaties.

Gaining compliance with vessel-source pollution standards has been a glaring weakness of the international effort thus far. Under the "flagship" principle, another of the basic tenets of customary ocean law, the only legal authority to which a merchant vessel is subject is the government of the state in which it is registered and thus whose flag it flies. Flag states assume the obligation to enforce international treaty rules for all ships under their registry when they voluntarily ratify such rules. Unfortunately, the pace of ratifications on IMO treaties has been woefully slow because maritime states are reluctant to put their fleets at a competitive disadvantage with those of countries that have not yet adopted them. Some states that have ratified the treaties are poorly prepared to enforce international rules on vessels that are dispersed over vast seas, while other states simply lack the inclination to take enforcement actions. A few states, known as "flags of convenience" set themselves up as havens from regulations, using lax rules and enforcement to attract the registration of foreign-owned vessels. The most notable examples of this type of free rider are Liberia and Panama, which between them account for the registry of a majority of the world's oil tankers, virtually all of which are owned by businesses in other countries.

The enforcement problem has received considerable attention over the past decade. At its 1973 conference, IMO debated a number of possible strategies for increasing the authority of coastal and port states over foreign vessels. However, in the face of opposition from maritime states that were concerned that any significant weakening of the flag of registry principle could lead to costly interference with international shipping, the issue was deferred to the United Nations Conference on the Law of the Sea, which began in 1974.[28] The new ocean treaty finally adopted in 1982 includes some important changes in regard to enforcement. Port states are given authority to prosecute operators of vessels stopping in their harbors that have violated international oil pollution standards *regardless* of where on the seas the infraction takes place. Port states are also permitted to make compliance with national or international safety standards a condition of gaining entry into their harbors. More limited powers are given to coastal states to set and enforce standards within their 12-mile territorial waters and 200-mile EEZ's. Normally, the procedure for coastal states will be to call infractions by foreign flag ships to the attention of the state of the next port of call, which is bound to take action against the violating vessel. These provisions and others will substantially erode the "floating sovereignty" of flag ships, which has been a major hindrance to the enforcement of international standards on marine pollution.

Case Study 4: Transboundary Acid Rain

Background

Wind currents carry polluting substances spewed into the atmosphere by one country into the airspace of others, where significant damage to the environment may occur. This phenomenon, known as transboundary air pollution, has become an increasingly serious international environmental problem over the past thirty years. Among the airborne pollutants that have evoked the greatest international concern are carbon dioxide, the buildup of which scientists fear is causing a warming of the atmosphere through what is called the greenhouse effect, something that may bring about a melting of the polar ice caps; chlorofluorocarbon (more popularly termed fluorocarbon) buildups, believed to be diminishing the protective stratospheric ozone layer, which filters out high levels of ultraviolet radiation from the sun that can cause skin cancer; and radioactive fallout, which results from the above-ground testing of nuclear explosives. In this case study we shall consider acid rain, a phenomenon that has drawn considerable international attention since its increasing spread and intensity was detected by a Swedish scientist, Svante Oden, just fifteen years ago.[29]

Acid rain is the result of a complex chemical change that occurs when oxides of sulfur and nitrogen are emitted into the air and are exposed to sunlight and water vapor. The resulting solution, which falls to the ground as rain or snow, is a diluted form of acid. It is also possible for the sulfuric and nitrogen oxides to form acid fogs or to settle to the earth in a dry form and become an acidic solution when they come into contact with surface waters. Some of the acid-forming compounds can be attributed to natural causes, such as volcanic activity. But beginning with the industrial age, this amount has been dwarfed by that which has been introduced by human activities. In modern times coal-fired power plants appear to be the greatest contributor of sulfur to the atmosphere. Other major sources are oil refineries, smelters, and various industrial operations. Petroleum-burning vehicles, especially the hordes of automobiles in industrial countries, are the source of most of the nitrogen oxides in the atmosphere.

Acid rain was almost exclusively a local problem until the relatively recent advent of high smokestacks, which were constructed in the belief that the harmful effects of pollutants could be diminished by dispersing them over a larger area. In the United States, 175 stacks that protrude five hundred or more feet into the sky have been built by utilities and industries to bring their operations into compliance with the Clean Air Amendments of 1970.[30] When introduced into the atmosphere at such high levels, the sulfur oxides may remain suspended in the air for days and even weeks, during which they can travel hundreds if not thousands of miles before precipitating to the ground. Thus, it is quite likely that much of the acid rain afflicting the eastern regions of the United States and Canada originates in the industrial belt of the American midwest, particularly the Ohio River Valley. Acid rain is an even more complicated legal and political problem in Europe, where numerous industrial states are found. Norway and Sweden have been especially concerned about the severe environmental damage attributable to acid-forming pollutants originating in the British Isles, France, Belgium, and the Germanies.

The acidity of water is measured using a scientific standard, the pH scale, which ranges from 0 to 14, with 7 being the neutral point between acidic and alkaline. Descending values correspond to a logarithmic increase in acidity. Thus, a pH value of 6 indicates that acidity is ten times greater than 7, a value of 5 is one hundred times greater than 7, and a value of 4 is one thousand times greater than 7. Rain having a pH value below 5.6 is considered abnormally acidic, and below 5.0 the acidic content is regarded as high. By the late 1970s the average pH level of rainfall in the eastern United States had fallen to below 4.5.[31] Some pH levels below 3, the acidity of vinegar, have occasionally been measured in Europe. In 1978 Wheeling, West Virginia, experienced precipitation having a pH of 1.5, which is more acidic than lemon juice.[32]

The most visible impact of acid rain is on the aquatic life of freshwater lakes, which is adversely affected when the pH level of the lake falls below 5.5 and virtually disappears as it descends to 4.5. In the case of fish, acidic waters not only deplete calcium from their bone structure, but also interfere with their reproductive processes. Furthermore, the acidity causes aluminum to be leached from the soil along with other heavy minerals; this clogs up gills, causing fish to die slowly and sink to the bottom of the lake. Much of the damage is done during the "acid shock" that takes place annually for a few days or weeks during the spring runoff when rivers and lakes may become as much as one hundred times more acidic than normal. In Ontario, Canada, 400 lakes have become biologically dead, while the aquatic life in 48,000 more is threatened.[33] In Sweden 4,000 lakes are already fishless and 14,000 more have been acidified to some degree.[34] It should be noted that the impact of acid rain varies from one region to another depending in large part on the presence (or absence) of limestone in the base rock, which will neutralize the acid.

Scientists are only beginning to understand some of the other consequences of acid rain. There is some evidence that the growth of some species of trees may be stunted by the acids, which could be a major blow to the lumber industries of a number of countries. Agricultural crops are affected, but here the findings are less consistent and conclusive. Acid rain damages the foliage of some crops and reduces yields. It can, however, have a fertilizing effect that increases the yields of certain grains, legumes, and fruits. Not to be overlooked is the impact that acid rain has in corroding man-made structures such as steel bridges and railroad tracks and the marble or limestone surfaces of buildings and monuments, including many of great historical significance that are adorned with elaborate stone sculptures. Acid also corrodes the insides of water pipes, contaminating the water they carry with heavy metals that are toxic to human beings.

Over the long run, acid rain may be a time bomb with much more serious environmental consequences than have been observed thus far. Unfortunately, it is likely that the sulfur emissions will continue to increase as coal is used to substitute for a risky reliance on high-priced petroleum from the Middle East.

Policy on Transboundary Air Pollution

Historically, the international community has done little to discourage transboundary air pollution. Even during the current era of heightened environmental awareness, international policy on acid rain and air pollution generally has been slower to unfold than has been the case with ocean pollution.

A small body of international case law suggests that countries can be held liable for damage they cause to the natural environment of other countries. Most directly relevant to transboundary acid rain is the often cited *Trail Smelter* case, which was heard by a special international tribunal. At issue was damage to the environment in the state of Washington from 1925 to 1937 caused by sulfur dioxide emitted from two 400-foot smokestacks at a large, privately owned smelter located across the border at Trail, British Columbia. In its decision, issued in 1941, the tribunal held that Canada was responsible for damages in the United States and for preventing harmful transboundary pollution from the Trail Smelter in the future. It concluded that "no State has the right to use or permit the use of its territory in such a manner as to cause injury by fumes in or to the territory of another or the properties or persons therein, when the case is of serious consequence and the injury is established by clear and convincing evidence."[35]

The responsibility of nation-states for the consequences that polluting activities have beyond their own borders was enunciated at the Stockholm Conference in 1972, which was originally proposed by Sweden as a forum in which concerns about acid rain could be discussed. Article 21 of the Declaration of Principles adopted at the conference provides that "states have . . . the responsibility to insure that activities within their jurisdiction or control do not cause damage to the environment of other states. . . ." The next article calls for the further development of international law regarding liability and compensation for such environmental damage when it occurs.

There is little in international treaty law that directly addresses the problem of transboundary air pollution. The outcry against radioactive fallout eventually led to the Limited Nuclear Test Ban Treaty of 1963, usually thought of as an arms control measure, which banned the testing of nuclear devices in the atmosphere, the oceans, and outer space. The only agreement that is directly applicable to acid rain is the 1979 Treaty on Long-Range Transboundary Air Pollution that was negotiated in the UN's Economic Commission for Europe (ECE). It was accepted by thirty-odd member countries among which are the United States, Soviet Union, and Canada, in addition to most of the countries of Eastern and Western Europe.

The ECE treaty obligates the signatories to limit and, as far as possible, to gradually reduce and prevent air pollution, especially sulfur dioxide emissions. Each country promises to use the "best available technology that is economically feasible" to combat air pollution, to cooperate on scientific research on the problem, and to exchange information on policies or undertakings that could cause significant long-term changes in transboundary pollution. A complaints procedure allows one country to call for consultations with other countries that it suspects are causing air pollution.

Although it is a significant first step in addressing the problem of air pollution, the treaty has been criticized for being vague and permissive, and its signatories have been slow to implement its provisions.

Why is international policy on air pollution still in an embryonic state? The relationship between polluting states and those that bear the consequences is often not reciprocal. States in which transboundary pollution originates have little incentive for bearing the economic cost of controlling emissions when the benefits would be enjoyed not by them but by downwind states. Part of the explanation lies in uncertainties about the seriousness of environmental harm from specific pollutants. Moreover, it is difficult to prove definitely that environmental damage occurring in one country is caused by pollution originating in another, given the variability of wind currents and the intermingling of pollutants. Even in the case of Canada and the United States, there is still considerable disagreement on the share of the blame for the sulfuric and nitrogen oxides that cause the acid rain that afflicts the two countries. Some of these deficiencies in our knowledge of transboundary pollution and its consequences will be remedied by continuing research efforts, including the efforts of UNEP through its Global Environmental Monitoring System (GEMS).

Summary and Conclusions

Each of the environmental problems described in the four case studies bears some similarity to the basic scenario of the tragedy of the commons. The closest match is perhaps with the depletion of ocean fisheries, where the fish and whales correspond to the grass of the pasture, the fishermen to the cattle, and the nations from which they come to the herdsmen. Discharging oil into the seas also has all of the basic elements of the tragedy if, rather than a pollution dump, the ocean is looked upon as a resource having a limited capacity, whose depletion will have undesirable consequences. The most notable differences between the tragedy of the commons and the case studies arise in the questions of the ownership of terrestrial species and the atmosphere. The notion that they are a common heritage and the property of humanity has yet to gain widespread acceptance in the way that the doctrine of the freedom of the seas, under which no national claims are recognized, has held sway since the time of Grotius. In all of the case studies, the benefits from using a resource accrue to the individual users—such as the fishermen, polluters, and poachers—and the costs from overuse are dispersed among the broader international community.

The most frequent response of the international community to an impending environmental tragedy of the commons has been to adopt restrictions designed to discourage overuse or misuse of the natural environment. Examples are the quotas of regional fishery commissions, the prohibition of trade in endangered species, and the rules of IMO on the discharge of oily substances into the oceans. However, the new Law of the Sea, with its provision for a 200-mile Exclusive Economic Zone, represents a major shift from international restrictions to national enclosures as a strategy for conserving fisheries. In the case of terrestrial species, the enclosures strategy implicit in national boundaries clearly has been inadequate, prompting international regulations as a supplementary protective measure. Acid rain is the one case where international policy has not progressed beyond voluntary restraint. More specific regulations may, however, be adopted in the not-too-distant future, given the increasing awareness of the magnitude of the problem. The international community is not ready nor does it have the capacity to use tax disincentives as a means of limiting consumption of scarce resources, even though they are a theoretical possibility in all four of the cases that were examined. Even less consideration has been given to setting up international public monopolies to achieve environmental objectives.

In each of the case studies, protection of the environment would appear to be the key objective of the global policies developed. Only in fisheries management schemes designed to achieve a maximum sustainable yield was production given similar weight as a policy goal. In regard to combating oil pollution and acid rain, the economic cost of control measures has seemingly been given less consideration, although many states have been slow to ratify agreements that might be costly. In the case of preserving terrestrial species, regulations have been made with care that trade in species that are not endangered not be unnecessarily restricted. Equity and freedom have been minimally involved as guides to policy. Expanding the exclusive jurisdiction of the coastal states over ocean resources is, for example, a substantial step away from equity in the distribution of the economic payoff derived from them, and significantly reduces the freedom of states with distant-water fishing fleets. States that are the exclusive host of exotic species have questioned the fairness of being asked to bear all of the costs of conserving them for the benefit of mankind; yet little assistance has been offered from the international community.

Substantial progress has been made in implementing global policies on ocean fisheries, terrestrial species, and oil pollution. In all three areas, agreement has been reached on elaborate treaties, and institutions have been established to provide for continuing implementation of the policies. The regional fishery commissions and IMO are most notable in this regard. Enforcement of rules among the parties to the treaties and coping with nonparticipating free riders represent continuing problems, especially in the case of conserving terrestrial species. The shift to an enclosures strategy for managing coastal fisheries may be successful in excluding free riders from localized stocks, but not for the highly migratory species. The new Law of the Sea sets forth more promising procedures for enforcing rules on vessel-source oil pollution. The response to the problem of transboundary acid rain lags far behind, the only accomplishment thus far being a vaguely worded regional agreement.

5 The Future of the Environment

In studying global environmental problems, our attention is inevitably drawn to the future. Most of us have adapted to the current condition of the environment and tolerate it even though it is far from being ideal. What is of greatest concern is the longer-term tendency for a rapidly growing human population to deplete and despoil nature. Younger and middle-aged generations living today may see a substantial deterioration of the environment during their lifetimes. Those who will probably be the most seriously affected are future generations that have not yet been born. In the past it was usually assumed that each generation left an enriched legacy for its descendants. Now it is not unlikely that our children and grandchildren will inherit a seriously overpopulated, resource-impoverished, and badly polluted world in which the quality of human existence has been significantly diminished.

In this chapter on environmental futures, we shall first discuss some of the different ways we think about and study the future. In particular, we will address the distinction between forecasting what is likely to happen as opposed to designing or planning futures that we would like to see evolve. Then, we shall consider a variety of environmental futures—both forecasts and designs—that have provoked considerable discussion in recent years.

Envisioning the Future

As a subject of study, the future differs from the past in some very fundamental ways. Whereas the past can be observed to the extent that there are adequate records of what has transpired, we can only imagine what the future will be like. And although students of history have only to examine a single chronology of events that have taken place, albeit a very complex one, as futurists we are confronted with an infinite number of possible futures from which to choose, ranging from the highly probable turn of events to the extremely unlikely.[1]

What we envision is a response to the questions we ask about the future. The most frequently asked question is what is the most probable future; for example, what will the weather be like tomorrow? Those futures believed to be most likely are known as forecasts. In thinking about the future, we may also ask which of the possible futures would be the most desirable and plan transitional strategies for bringing the most desirable possibility into being. Thus, to extend the example of future weather conditions, we may decide upon the type of weather we would like to have, such as adequate rainfall for agricultural operations, and use techniques such as cloud seeding to bring them about. The futures we desire are often referred to as designs for the future, or when they are fanciful or appear unattainable, they are described as utopias. Thus in contrast to forecasters, who normally assume the passive posture of a spectator of the future, designers often entertain the notion of active interventions to steer the course of future developments in a desired direction.[2]

Forecasting the Probable

Forecasts have been defined above as visions of probable futures. In many cases, forecasts presume the continuation of certain trends and conditions. Thus, the envisioned future is forecast to be the most likely outcome only if those conditions are present. In the environmental field, for example, forecasts are often made with the proviso that present trends will continue through the period, even though this may not be likely or even possible. Likewise, a forecast may be predicated on the absence of any major surprise developments, such as extreme changes in climate or the outbreak of all-out nuclear war between the United States and the Soviet Union, either of which would have a decisive impact on the future. Sometimes forecasts may be predicated on expected policy changes, such as the tightening of rules limiting emissions of pollutants.

The methods used in forecasting range from prophecies based on superstition, such as those of palm readers or astrologers, to the considered judgments of well-informed experts or more quantitative projects of scientists. Most serious forecasting is based upon information and knowledge of the past. The most conservative forecasts suggest that nothing significant will change and, thus, that the future will be much the same as the past. Obviously, such a forecast is more likely to be valid over the short run than the long term, during which there is a higher likelihood that unforeseen events will take place. Many forecasters, especially in the environmental field, look to the past for trends, and project them into the future on the assumption that the rates of change will remain constant. Inevitably trends will undergo change when a critical threshhold is reached, the timing of which is often difficult to anticipate. Some forecasters attempt to take into account the ways in which trends interact with one another; for example, the extent to which fertility rates are related to economic progress in the less developed countries. The most sophisticated of these forecasts, discussed later in this chapter, are based on complicated computer simulation models, which plot the interactions between hundreds of variables using thousands of equations.

Not all forecasting involves trend projection. For example, if we wish to forecast what will follow from a current set of conditions, we might look back to a similar set of circumstances in history and, on the basis of what happened then, predict future developments. In many cases, however, it is difficult if not impossible to identify a historical context that is sufficiently parallel to the present to yield reliable forecasts.[3]

It is tempting to judge the quality of a forecast by whether it comes true, that is, whether future events end up being similar to those that were foreseen. This is not always, however, an appropriate way to evaluate forecasts. It can happen that a good forecast—one that effectively takes into account all available information—may itself become a factor influencing the future. This can occur when a forecast of especially undesirable events attracts widespread attention and provokes a public effort to avert it. To some extent this has been the case with the alarming forecasts advanced in the 1972 Club of Rome report, discussed later in this chapter.[4]

Designs of Preferred Futures

The impetus for designing preferred futures usually results from an awareness that the future that is the most probable is quite different from what we would like it to be. A catastrophic future that is less probable, but yet possible, may also trigger efforts to ensure that such an eventuality will be averted. If the probable future is a desirable one, or at least no worse than current conditions, there is little reason for contemplating an alternative to it.

Ideally, the designer of preferred worlds envisions a wide range of alternatives that would each be an improvement over where society appears to be headed. Designing usually requires more creativity than making forecasts, which often involves a simple projection of current trends. The inspiration for a preferred future often is found in a problem-solving strategy that has been successful in the past, in another contemporary setting, or on a smaller scale. It may also be based on a logical deduction of what the designer calculates should work. Or it may be created by tinkering with projected trends. Thus, as with forecasts, some designs are a product of personal intuition, others of elaborate computer simulation models.[5] The preferred future is chosen from other futures on the basis of both its intrinsic desirability and its practicality. Having decided on a preferred future world, it behooves the designer to demonstrate that it can be brought into existence, and that it is not simply a fanciful utopia.

The odds against successfully designing and implementing a preferred future are great, especially one having an international or global scope. Any effort to bring about major social changes will be resisted, sometimes strenuously, by those who have a stake in the status quo. Opposition may also come from those who fail to recognize the severity of looming problems to which a response is imperative, from those who disagree on what the most preferable future should be, or from those who argue for a different strategy to bring it into existence. Even if a preferred future has widespread acceptance, it may not evolve in the way that is planned, either because of miscalculations on the part of the designer, a failure to take complicating factors into account, or unanticipated changes in the circumstances upon which the design was predicated.

In spite of these obstacles, there is no dearth of designs of preferred futures, and especially of designs having global scope. The Council on Foreign Affairs, in its 1980s Project, sponsored a series of designs of international futures, each of which addressed specific problems as diverse as the world economy, monetary disorder, nuclear weapons, human rights, and disaster relief. The Institute for World Order, in its World Order Models Project (WOMP), has undertaken an especially ambitious project to design a series of preferred futures for the 1990s. These designs, each of which is written from the perspective of a different region of the world, attempt simultaneously to address a full array of the most serious global problems.

Global Environmental Forecasts

Forecasts of the future dealing in whole or in large part with global environmental matters range from confident optimism to extreme pessimism. Over the past decade or two, there has been a spate of ominous warnings of difficult if not apocalyptic times ahead for humanity if current ecologically related trends continue indefinitely into the future. What makes the future look even more threatening is the way that environmental problems are interrelated, with one trend reinforcing others; for example, population growth speeds up resource use, which in turn results in more pollution and ravaged landscapes. Inequities in the distribution of wealth and the massive buildup of arms by sovereign nation-states further complicate the situation. The concept "global problematique" has been applied to this group of interrelated problems, or to state it more dramatically, this "crisis of crises."

The doomsday scenarios have been challenged by critics who argue that their authors are pessimists whose forecasts of overpopulation, scarcity of resources, and environmental degradation will prove to be no more valid than the discredited projections of overpopulation and food scarcities made by Thomas Malthus nearly two hundred years ago. The most extreme among the optimists predict that human beings will transcend what are now natural limitations through technologies that will permit continuous and perhaps even spectacular improvements in life-styles. In the sections that follow we shall consider several prominent examples of both the pessimistic and optimistic forecasts.

Pessimistic Forecasts

Among the pessimistic forecasts, perhaps the most widely read and influential one has been *The Limits to Growth* (1972), a report prepared by a research team led by Dennis and Donella Meadows at the Massachusetts Institute of Technology. The report was commissioned by the Club of Rome, a group of prominent businessmen and academicians from twenty-five countries who were concerned that the major problems confronting humanity were so complex and interrelated that traditional institutions and policies would not be effective in coping with them.[6]

The MIT group tackled the question of how long growth in population and industrial production could continue, given what was known about the maximum amount of food that could be produced globally, what remains of energy and mineral resources, and the capacity of the ecosphere to absorb pollutants. A computer simulation model was used to project these trends through the twenty-first century taking into account the natural limitations that could affect them. The principal conclusion of the study was that a continuation of current growth trends in world population, industrialization, pollution, food production, and resource depletion would, within the next hundred years, lead to a situation in which growth would no longer be possible and, worse yet, to "a rather sudden and uncontrollable decline in both population and industrial capacity."[7] Figure 5.1 portrays a more specific scenario of what the Meadows team calculated would happen if there were no major changes in the physical, economic, or social relationships that have prevailed historically.

What makes the forecasts contained in *The Limits to Growth* even more discouraging is the likelihood of a collapse even if some unrealistically optimistic assumptions are made. Rerunning the

Figure 5.1 World model standard run. Source: Donella Meadows et al., *The Limits to Growth* (New York: Universe Books, 1972), p. 124.

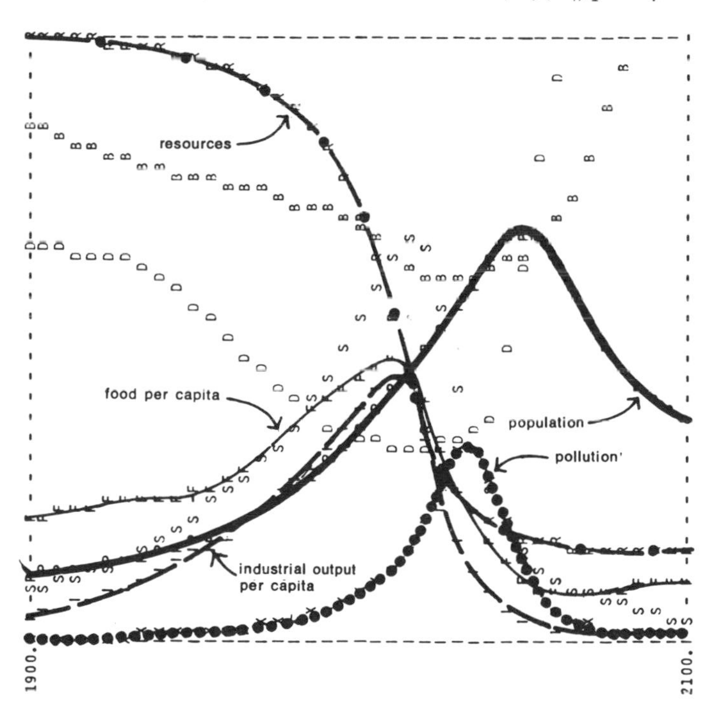

simulation model with a favorable assumption that resource reserves are twice today's known levels results in a more rapid increase in pollution, which in turn causes a sharp decline in food production and an increase in the death rate.[8] Adding an even more optimistic prognosis of unlimited resources and effective pollution controls makes more growth in population and industrial production possible, but this causes a reduction in the amount of arable land available for agriculture and therefore a reduction in the amount of food grown per capita. Industrial growth then slows as capital is diverted to food production.[9] Adding the assumption that food production per acre can be doubled or that a dramatic increase takes place in the effectiveness of birth control programs does little to postpone the date of ecological reckoning.[10] Perhaps most significant, however, is that no allowances were made in any of the runs for political or economic problems, such as wars, depressions, embargoes, and gross inequalities, that would result in a less-than-optimal use of remaining natural resources. These findings were shocking to a world that was accustomed to taking growth for granted. The influence of the report was especially great because it appeared to have the authority of modern computer science behind it in contrast to other warnings of ecological doom coming from individual scientists.

The Club of Rome commissioned a second study that was undertaken by an international team of specialists led by Mihajlo Mesarovic from the United States and Eduard Pestel from West Germany. Their report, *Mankind at the Turning Point* (1974) did not, however, have as great a public impact as did its predecessor, *The Limits to Growth*. The Mesarovic and Pestel team also based its forecasts on the projections of a simulation model, but one that was far more complex, most notably in dividing the world into ten regions, each of which has its own unique set of ecological circumstances. Rather than forecasting a general global collapse of population and economic growth tendencies, as the Meadows team had done, *Mankind at the Turning Point* warned that "catastrophies or collapses on a regional level could occur, possibly long before the middle of the next century, although in different regions, for different reasons, and at different times."[11] The situation in the less developed regions was seen to be especially ominous because of projected rapid population growth. By the year 2000 each square kilometer of cultivated land in South Asia would have to feed 390 additional people, compared to only 37 more people in North America.[12] The people of the developed world should not become complacent, however, because the world is an interrelated system and, thus, regional catastrophies "will be felt profoundly throughout the entire world."[13]

A more recent addition to the pessimistic forecasts of environmental futures is the *Global 2000 Report to the President* (1980), which was based on trend projections using the data and the modeling capabilities of a number of federal agencies. The principle conclusion of the study was the following forecast:

> If present trends continue, the world in 2000 will be more crowded, more polluted, less stable ecologically, and more vulnerable to disruption than the world we live in now. Serious stresses involving population, resources, and environment are clearly visible ahead. Despite greater material output, the world's people will be poorer in many ways than they are today.
>
> For hundreds of millions of the desperately poor, the outlook for food and other necessities of life will be no better. For many it will be worse. Barring revolutionary advances in technology, life for most people on earth will be more precarious in 2000 than it is now—unless the nations of the world act decisively to alter current trends.[14]

In addition to the trends in population, food, resources, and pollution that were featured in the reports to the Club of Rome, *The Global 2000 Report* looks at water shortages, the loss of forests, the deterioration of agricultural lands, and extinctions of plant and animal species. The projections are based on the assumption that existing national policies on population stabilization, resource conservation, and environmental protection will remain unchanged through the end of the century.[15] It also assumes that rates of technological development will continue and that no major political or economic disturbances will take place.

It should be kept in mind that the objective of the authors of these doomsday forecasts was not to predict what the future *will* be like, but to call attention to the undesirable consequences if present trends were to be allowed to continue indefinitely. Their hope was that this knowledge would spur the adoption of policies that would lead to a more harmonious relationship of humanity with nature. Both of the Club of Rome models were used to demonstrate how this could be done by exploring possible futures in which growth was limited by corrective policies to a point where an ecological equilibrium was achieved.

Optimistic Forecasts

Not all prognosticators of global ecological futures accept the foreboding scenarios described in the previous section. The two reports to the Club of Rome have come under sharp attack from specialists in a variety of fields who challenge both the assumptions and data that were used in deriving the ominous projections. Especially strong criticisms were leveled at *The Limits to Growth*, which was the simpler and more widely read of the two.

Several of the critics are much more optimistic about the adequacy of the resource base of the planet for sustaining economic growth well into the future. English economist Wilfred Beckerman points out that the Meadows forecasts are only the latest of a series of predictions over the past two centuries of impending resource scarcities that would limit economic growth, all of which have been proven by history to be overly pessimistic.[16] Among these was the English coal scare triggered by W. Stanley Jervons's warnings in 1866 that the imminent exhaustion of England's coal supply would cause its industry to grind to a halt. Even with much more of an increase in the demand for coal than he anticipated, known reserves of coal have since mushroomed to an estimated 600-year supply.[17]

Beckerman contends that the Meadows team repeats some of the same mistakes that caused Jervons and other forecasters of approaching resource scarcities to be far off the mark. The most serious of these errors is to assume that statistics on reserves of natural resources are a valid indication of how much of them will ultimately be available, and therefore of how many years of current consumption they will sustain. When reserves are sufficiently large to satisfy projected demand, there is little incentive to undertake what can be costly exploration for new reserves. When they drop below this level, companies search for additional supplies and the known reserves are built up again. Moreover, as temporary scarcities drive market prices higher, it becomes economically profitable to make use of less accessible resources that previously would have been too costly, as in the case of drilling for offshore oil at greater and greater depth. Higher prices also spur efforts to find substitutes for the scarce resources or ways to conserve or recycle them. Higher prices also encourage investment in the technologies that would make these options feasible.[18]

The doomsday forecasts have also been criticized for being based on a misreading of historical trends. Resource economist Julian Simon's recent book, *The Ultimate Resource*, which has provoked much heated discussion, cites a number of more encouraging trends that lead him to believe that the human race will be no worse off in the future than in the past. He notes that declines in the rate of population growth have caused the United Nations to revise its earlier forecasts of the world's population in the year 2000 downward from 7.5 billion (1970 forecast) to 5.6 billion (1975 forecast).[19] The availability of land, he contends, poses no foreseeable constraint on future increases in production. Simon suggests that energy and other natural resources are actually becoming less scarce, if scarcity is measured in terms of the price of them relative to wages.[20] The sharp increases in the price of petroleum since 1973 are attributable to political maneuvering rather than to scarcity of reserves.[21] Pollution, at least of the types that are harmful to human health, he argues, is diminishing. His evidence is an increase in life expectancy accompanied by a trend away from deaths caused by environmental factors to afflictions that naturally occur in old age, such as heart disease, cancer, and strokes. Contrary to popular belief, upward tendencies in cancer rates, he believes, are due not as much to carcinogens in the environment but to more people living to the cancer-prone ages.[22]

The most optimistic projections are perhaps those of Herman Kahn and his associates in *The Next 200 Years*. Kahn views the 1970s as the midpoint in a 400-year transition from a condition in which all societies were preindustrial to one in which virtually all will have achieved a postindustrial state. On the basis of what he considers to be a relatively conservative, "earth-centered" perspective, Kahn forecasts that by 2176 world population will have flattened out at 15 billion, the annual per capita product will be $20,000, and the gross world product will be about $300 trillion.[23] On the subject of growth he maintains that

> current trends point to the conclusion that growth is likely to continue for many generations, though at gradually decreasing rates, which we expect to result more from a slowing pace of demand than from increasing difficulties in obtaining physical supplies. According to this analysis, the gradual leveling-off tendency will be a social consequence of the proliferation of such factors as modernization, literacy, urbanization, affluence, safety, good health and birth control, and governmental and private policies reflecting changing values and priorities. . . . Although the possibility of overcrowding, famine, resource scarcity, pollution and poverty cannot be dismissed, they should be seen as tempo-

> rary or regional phenomena that society must deal with rather than as the inevitable fate of man.[24]

Underlying Kahn's optimism is the conviction that the critical resources are not the natural ones, but are instead capital, technology, and educated people for which there are no meaningful limits.[25] Kahn also offers a second, even more optimistic "space bound" scenario that assumes a vigorous attempt to make use of outer space in the early twenty-first century.

The forecasts of the optimists have also come under intense attack. Simon has been criticized for a selective and manipulative use of fragmented statistics to support his optimistic outlook. For example, in noting that food production has kept pace with population growth over the past several decades, he fails to mention that the rate of increase has steadily declined, a trend that does not bode well for the future. Likewise, Simon's analysis of trends in pollution is limited to statistics from the United States and to the two of five major types of pollution for which declines have been recorded.[26] Of Kahn's forecasts, it could be said that he unrealistically assumes the inverse of Murphy's law: "If everything can go right, it will." The solutions he foresees to the global problematique are likely to be sidetracked by shortages of several key resources, the unavailability of sufficient investment capital, severe environmental impacts, and political and social complications.

Global Environmental Designs

Forecasts of ecological catastrophe have spawned many proposals for what humanity should do to avert such an eventuality. These designs form a diverse group. Some focus very narrowly on specific ecological problems. At the other extreme, some designers, impressed with how environmental problems are interrelated both among themselves and with other global problems, have set forth comprehensive strategies for coping with a troubled future. Some designers look to new technologies as a way of coping with environmental problems, others to economic policies or new political institutions as the most promising way to bring about needed changes. Finally, some plans for the future recommend only relatively minor modifications in existing policies and institutions, while others stress the need for fundamental structural changes in the existing order.

In this section we shall consider a variety of approaches to designing an alternative future that copes with the global ecological problems that are emerging. These approaches are (1) international regimes, (2) steady-state economics, (3) centralized political authority, (4) local self-reliance, (5) global equity, and (6) high technology.

International Regimes: The UNEP Approach

The proponents of the international regime approach to designing a future world order look favorably on the way the international community has been addressing a number of global problems, including what has been done on environmental matters over the past decade with the establishment of the United Nations Environment Program, and the continuing work of a number of specialized international governmental organizations (IGOs). The combination of the IGOs and the international policies that address any given policy problem are now frequently being referred to as an "international regime." Thus, for example, the International Whaling Commission and its policies are the core of a regime designed to preserve endangered species of whales. In assessing the accomplishments of these institutions and policies, the supporters of international regimes view the proverbial water glass as being half full rather than half empty, inasmuch as significant progress has been made in the right direction, but much remains to be done to complete the task at hand.

No comprehensive world regime is envisioned for the future, but rather a continued growth in the loosely coordinated networks of IGOs that address a limited range of environmental problems. Regional IGOs with a narrower geographical focus would also proliferate where environmental problems of common interest are more localized. In addition, new international treaties and assistance programs would be established, existing policies would be refined, and more effective techniques for enforcing international regulations would be devised. The result would be the evolution of increasingly elaborate international regimes, which would more successfully tackle the global agenda of environmental problems. Such a possibility bears some notable resemblances to the "functionalist" theory of a generation ago, which suggested that the most promising approach to international peace lay not in unlikely-to-succeed attempts to establish a world government, but in binding nations together in a growing network of IGOs having specific tasks that they could successfully perform. Their successes would then serve as a model for international cooperation on other problems.

The regime approach presumes that national governments are willing to cooperate with one another and, in so doing, to sacrifice a substantial part of the sovereignty that they have traditionally guarded so jealously. What prospect is there that they will be willing to do this? The hope lies in the recognition by national leaders of how interdependent their countries have become in environmental matters and how their interests here as well as elsewhere are served by stronger international regimes that can effectively manage this increasing interdependence. Fully understanding this, national governments may be willing to compromise on their own immediate interest in order to do their part to establish international regimes that will serve their own long-term goals and those of humanity as well.

The Steady-State: from Cowboy to Spaceman Economics

Warnings of impending resource scarcities have stimulated a reevaluation of long-standing and widely accepted theories of economics oriented toward maximizing growth. Steady-state economics has been proposed as an alternative paradigm that is more in tune with contemporary ecological realities. Economist Kenneth Boulding describes the transition that humanity must make in metaphorical terms: from a "cowboy" to a "spaceman" economy. The cowboy is symbolic of "illimitable plains and of reckless, exploitative, romantic, and violent behavior, which is characteristic of open systems."[27] The wide-open spaces over which the cowboy roams have resources in sufficient abundance to satisfy the desires of the sparce frontier population, and what seems to be a limitless capacity to absorb its pollutants. Under these circumstances it seems appropriate to exploit greedily what the land has to offer without concern for conservation of resources or the quality of the environment. In contrast, the spaceman lives in a closed system, a spaceship, which offers its inhabitants cramped quarters. They must survive on the limited provisions with which the spaceship is stocked and must carefully dispose of their wastes so as not to poison their small environment. These circumstances dictate a slow, planned rate of consumption that not only conserves supplies, but also minimizes the amount of wastes that are generated.[28]

The intellectual father of steady-state economics is John Stuart Mill who wrote more than a century ago: "I cannot . . . regard the stationary state of capital and wealth with the unaffected aversion so generally manifested towards it by political economists of the old school. I am inclined to believe that it would be, on the whole, a very considerable improvement on our present condition."[29]

Mill also observed:

> A population may be too crowded, though all be amply supplied with food and raiment. It is not good for a man to be kept perforce at all times in the presence of his species. . . . Nor is there much satisfaction in contemplating the world with nothing left to the spontaneous activity of nature; with every rod of land brought into cultivation, which is capable of growing food for human beings; every flowery waste or natural pasture plowed up, all quadrupeds or birds which are not domesticated for man's use exterminated as his rivals for food, every hedgerow or superfluous tree rooted out, and scarcely a place left where a wild rub or flower could grow without being eradicated as a weed in the name of improved agriculture. If the earth must lose that great portion of its pleasantness which it owes to things that the unlimited increase of wealth and population would extirpate from it, for the mere purpose of enabling it to support a larger but not happier or a better population, I sincerely hope, for the sake of posterity, that they will be content to be stationary, long before necessity compels them to it.[30]

Following in the same school of thought, Herman Daly, one of the leading contemporary advocates of steady-state economics, suggests that an emphasis on growth fails to take into account two basic laws of economics. The first is the law of "diminishing marginal utility," which suggests that people satisfy their most pressing needs first. As they consume more and more, the needs they satisfy are less important to them. The second is the law of "increasing marginal cost," which implies that what is produced first is usually made from the resources that are most readily available and the least costly to use in both economic and environmental terms. As more and more is produced, it becomes necessary to tap less accessible resources and, in so doing, to cause more environmental devastation.[31] In general, the developed countries have progressed much further than has the Third World toward the point at which the value of increased consumption is no more than the costs of increased production. That point may in fact have been passed some time ago by many of the developed countries.

The designs for steady-state economies would postpone the time of ecological reckoning by switching the emphasis from the current economic goal of maximizing extraction, production, and consumption, to a goal of maintaining a certain material standard of living. While that standard could be achieved by making available large quantities of short-lived products, as is so often done in our contemporary throwaway society with its disposable, nonrepairable products, steady-state economics would achieve it with a much smaller stock of durable items that do not require frequent replacement. The result would be a more frugal use of scarce resources with a less damaging impact on the environment. Recycling is another steady-state possibility so long as the process is not too costly to be economically competitive, is not a heavy consumer of scarce resources such as energy, and is not itself a cause of serious environmental pollution.[32] Usually recycling is less costly in these terms than mining and processing virgin materials.

Significant social issues arise in conjunction with steady-state solutions to global ecological problems, especially in regard to the distribution of wealth. Growth has traditionally been a means to placate those at the bottom of the socioeconomic scale of society who can be made content with some improvement in their living standard, even if their minimal share of the total wealth remains the same or even declines. With growth, everyone will get more no matter how little. Without growth, the have-nots feel trapped in their condition and become resentful of inequities. They may react by challenging the prevailing order. Similarly, in the international arena there is concern by the Third World that the reduction of rates of growth by the adoption of steady-state strategies in the developed world would slacken demand for their exports and, in so doing, would thwart their own growth plans and would freeze the gross inequities in the distribution of the world's wealth that now prevails. Within societies, a slowing of growth could also exacerbate unemployment problems as fewer workers would be needed to produce a lesser quantity of goods. As Daly expresses it, under the prevailing economic system, to earn an income we "produce junk and cajole other people into buying it."[33]

These concerns have been addressed by steady-state economists in two ways. First, not all growth would be limited. Growth would only be limited if it requires the use of scarce natural resources or has adverse environmental consequences. Opportunities might grow in the service sector or in the areas of art and culture. Second, some redistribution of income may be necessary and have desirable consequences. For example, each of the limited number of jobs might be shared by more than one person, which would then make significantly more time available for leisure pursuits. Added leisure time would also invigorate the service sector.

Centralized Political Control: An Authoritarian World Government

Another school of thought argues that our popular beliefs about political authority are reflections of the historical context from which they emerged. The ideas of democracy and a limited government that maximizes individual freedoms, emerging from seventeenth and eighteenth century political thought, became fashionable when resources were much more plentiful in relation to population than is now the case for most countries. Laissez-faire policies were especially well suited to the American scene a century or two ago, when a very small population was just beginning to explore and exploit the continent's vast storehouse of resources.[34] The philosophical base for American democracy supports individual rights and individual acquisitiveness ("life, liberty, and the pursuit of happiness").

The circumstances in which humanity now finds itself have changed dramatically. Freedom to exploit nature's bounty combined with decades of historically unprecedented rates of population growth have sharply altered the resource situation from plenty to growing scarcity. Conflict over the remaining resources may become more intense, with scarcity increasing the potential for political instability within countries and warfare between them. In the end, freedom and a weak political authority bring about "total ruin," to borrow a phrase from Garrett Hardin.[35] Several theorists have argued that a sacrifice of some freedoms will be necessary to avoid ruin. The role of government would then change from protecting individual rights in order to enhance individual acquisitiveness, to protecting the common interests of the majority in order to preserve resources for all. This, of course, is one of Hardin's preferred strategies for averting a tragedy of the commons—"mutual coercion, mutually agreed upon by the majority of the people affected."[36]

The case for a strong, centralized authority to enable humanity to cope with its global ecological problematique has been made most forcefully by Robert Heilbroner and William Ophuls. Heilbroner looks to national governments to exercise control over their own populations in order to curb some of the excesses that are bringing humanity to the brink of ecological collapse. In his widely read book *Inquiry into the Human Prospect*, he suggests that "the passage through the gauntlet ahead may be possible only under governments capable of rallying obedience more effectively than would be possible in a democratic society."[37] Thus, for example, it could be argued that it will be necessary for the Indian government to adopt a more coercive strategy of population control if that country is to avert an ecological catastrophe in the not-too-distant future. Similarly, Ophuls in *Ecology and the Politics of Scarcity* argues that a high level of freedom is no longer appropriate, given current ecological realities: "Accordingly, the individualistic basis of society, the concept of inalienable rights, the purely self-defined pursuit of happiness, liberty as maximum freedom of action, and laissez faire itself all become problematic, requiring major modification or perhaps even abandonment if we wish to avert inexorable environmental degradation and eventual extinction as a civilization."[38]

Even more problematical is the freedom that nation-states exercise in accordance with the long-standing principle of sovereignty. The result is a highly fragmented political order with nation-states competing for resources. The authoritarian solution is a highly centralized world government, as Ophuls suggests:

> Thus, the already strong rationale for a world government with enough coercive power over fractious nation states to achieve what reasonable men would regard as the planetary common interest has become overwhelming. Yet we must recognize that the very ecological scarcity that makes a world government ever more necessary has also made it much more difficult of achievement. The clear danger is that, instead of promoting world cooperation, ecological scarcity will simply intensify the Hobbesian war of all against all and cause armed peace to be replaced by overt international strife.[39]

Local Self-Reliance: Small Is Beautiful

If there is an alternative to an authoritarian world government, perhaps it can be found in the title of E. F. Schumacher's widely read and discussed book, *Small Is Beautiful*.[40] Schumacher and his intellectual disciples have argued that the ecological crisis we face is at least in part attributable to the large-scale modern industrial enterprises and the tendency for industrial societies to become increasingly concentrated in sprawling urban areas. They suggest that a solution to the crisis may lie in scaling down these enterprises and reducing the size of human communities to the point where they can become largely self-sustaining units that have a much smaller impact on the environment. The message here is also applicable to Third World countries where urban areas have been mushrooming as they have attempted to develop according to the Western model of industrialization.

Many features of modern urbanized industrial society can be indicted on ecological grounds. The inhabitants of large cities rely upon food brought in from the outside. In some cases the food is transported thousands of miles from where it is grown on big, highly mechanized farms that consume prodigious quantities of fossil fuels and apply heavy doses of fertilizers and pesticides that pollute the environment. As the size of cities increases, their residents must commute longer distances, thus increasing the consumption of gasoline and emitting greater quantities of noxious pollutants in the air. Urban areas have vast needs for electricity, most of which is generated by burning increasingly scarce petroleum or more plentiful coal, something that can cause significant damage to the environment. A strong central authority is needed in such a complex, interdependent society to prevent breakdowns in the supply networks upon which the urbanities depend. Moreover, the residents of these concrete and glass environments, with their highly specialized occupational niches, almost completely lose contact with the natural world that is essential to their survival, and therefore have little appreciation of what is necessary to preserve it.

The "small is beautiful" alternative encourages the development of decentralized urban neighborhoods and a movement of people back to small rural communities in which they would live more in tune with the environment. They would make use of what are called "intermediate," "appropriate," or "alternative" technologies that can be adopted on a small scale and that make use of renewable resources. Residential energy needs, such as for heating space and water, would be satisfied by solar collectors. Residents would be geographically close to work-places, schools, shops, and other services, thus reducing the need for mechanized means of transportation, which consume great quantities of imported petroleum and pollute the atmosphere. Most of the food they consume would be grown organically, thus avoiding the environmental pollution caused by the runoff of chemical pesticides and fertilizers. Sewage and other organic matter would be composted for fertilizer. Lumber used in construction would be harvested from local woodlots. The inhabitants of these largely self-reliant communities would be constantly reminded of their dependence on nature and consequently be more committed to preserving it. In this way, the "small is beautiful" approach incorporates Hardin's "intrinsic responsibility" as an incentive for environmental stewardship.

These smaller-scale communities would also foster very different types of social relationships. The familiarity that grows out of frequent face-to-face contact would lead to an enriched social environment in which its members would offer each other support and assistance, such as in community house raisings. Plumbers, electricians, and dentists might offer their services to one another on a barter basis. The relative simplicity and self-reliance of the smaller communities would reduce the need for a complex social hierarchy, over which the people can exercise little control. The governance of the communities could be truly democratic, perhaps along the lines of the town meetings of New England villages.

Could we go back to smaller-sized communities after having invested so heavily in specialized occupations and in the centralized infrastructure and supply networks of urban areas? Interestingly, the 1980 census showed that more people moved from the cities to the countryside during the past decade than the reverse. In addition, a number of communities have been established, among which are the New Alchemy Institute in Massachusetts and the Meadowcreek Project in Arkansas, which demonstrate alternative, ecologically responsible life-styles.[41] Then there are the Amish who in their small communities dotting parts of Canada and the United States have practiced the "small is beautiful" way of life for centuries, while shunning the conveniences of the industrializing society around them.[42] Although it is doubtful that a very sizable proportion of the population could make such a complete transition to an alternative life style within the foreseeable future, enough people may adopt whatever elements of it that they can—such as by the use of solar power in their homes—to make a noticeable difference in the overall resource consumption of their societies.

Global Equity: A Compact with the Third World

Some designers argue that gross inequities in the distribution of wealth between North and South must not be ignored in efforts to address global ecological problems. Not only is it ethically repugnant for a minority of the world's population living in the developed regions to consume wastefully a lion's share of what have become scarce natural resources while introducing a disproportionate share of pollutants into the environment, but in an increasingly interdependent world, the "haves" of the North must also rely more and more on the goodwill and cooperation of the South in addressing global problems. Over the past decade the Third World countries, coalescing as the Group of 77 and the Nonaligned States, have repeatedly made it clear at world conferences that their willingness and capacity to address the problems of special concern to the North will be conditioned on the commitment of the North to assist them in achieving their development objectives.

The interdependence of North and South is especially pronounced in environmental matters. With mushrooming populations and plans for rapid economic development, the Third World countries could significantly damage the planetary ecosystem in ways that would adversely affect the developed world. Of particular concern are the burning of cheap, high-sulfur coal that would contribute to acid rain problems and intensify the greenhouse effect; the widespread use of durable pesticides such as DDT that gradually work their way to even the remotest areas of the planet; massive land-clearing projects that destroy the habitat of endangered species; and the dumping of sewage and industrial wastes into the oceans and regional seas. In many cases, the less developed countries lack the technology and capital to minimize the environmental impact of their development programs. The developed world should also not overlook its growing reliance on the Third World for resources after having plundered its own resource endowment.

Recognizing their dependence on the Third World, the developed countries may find it to their advantage to enter into an understanding with them, a kind of "global compact" or "planetary bargain," which incorporates trade-offs on several major issues, including concessions on the international economic order that would give the Third World a greater share of the world's wealth.[43] One possibility would be to start implementing the New International Economic Order (NIEO), which the Third World has been calling for over the past decade.

Several provisions of the NIEO are of relevance to the environment. One is the principle that all countries are entitled to exercise sovereignty over the natural resources within their boundaries regardless of what concessions were made to foreign interests in the past. Others would stabilize the prices of the primary commodities, including unprocessed agricultural products and minerals, upon which many less developed countries heavily rely for export revenues. This would avoid the sharp price fluctuations that have regularly occurred in the past and that have made long-term economic planning so difficult. The prices of these primary commodities would also be "indexed" to rise at the same rate as the prices of the industrial goods that Third World countries must import, and that historically have gone up much more rapidly. Finally, it has been proposed that the proceeds from the development of the resources of the international commons, including the oceans, seabed, and outer space, could be used to provide a continuing source of development assistance that would be independent of the political uncertainties of national governments. In return, the developed countries might be guaranteed certain levels of imports of needed raw materials.

Numerous other proposals have been made. Reacting to *The Limits to Growth*, a group of Latin American scholars developed their own simulation model, which was used to explore possibilities for a new international order called an "egalitarian society" that would be more humane than the existing one and at the same time more compatible with the environment. A new pattern of ownership (which would essentially be socialistic production) would orient society toward satisfying basic human needs, rather than generating profits. A roughly equal division of consumption among four regions—Latin America, Asia, Africa, and the "rich" countries—would eventually be achieved, in part as a result of having the rich countries transfer 2 percent of their incomes to the regions of Africa and Asia.[44] In a scenario of what could take place up through 2024, McGeorge Bundy envisions a world in which the superpowers lay down their arms after a series of nuclear "events" and allocate the savings to an assistance program that guarantees sufficient food to feed the Third World peoples in return for their commitment not to acquire armaments and to stabilize their populations at agreed-upon levels.[45]

Would the developed countries be buying peace with the Third World at too high a price as they transfer a significant amount of their wealth to the world's have-nots? Advocates of the global equity future contend that this would not be the case. Rather, the growth of the economic systems of Third World countries would result in greater demand for the goods produced in the industrialized nations and thus contribute to greater prosperity throughout the world.

High Technology: Terrestrial and Extraterrestrial Possibilities

Technological optimists contend that it is unnecessary to resign ourselves to limiting population growth, steady-state economic systems, simpler life-styles, or more coercive political systems. The answers to the ecological problems that loom ahead are not to be found in devising plans for living within what are currently perceived to be the natural limits of our environment, but rather in transcending them by unleashing scientists and engineers to achieve technological breakthroughs that will make continuing growth possible indefinitely in the future. Two of the most dramatic frontiers of high technology that could have a bearing on global ecology are genetic engineering, which among its many possibilities has the potential for spectacular increases in food production, and space colonization, by which man could tap the virtually limitless resources of outer space.

In the field of agriculture, there is much hope that genetic engineering involving recombinant DNA will make a second "green revolution" possible, enabling a much greater increase in food production than that resulting from the first green revolution of the past two decades (which developed and introduced into widespread use high-yield strains of wheat and rice). One disadvantage of the "miracle" crops of the first green revolution, developed by traditional techniques of cross-fertilization among different strains of related species, has been the large quantity of water, fertilizers, and pesticides needed to grow them. Genetic engineering makes it possible to crossbreed species from different evolutionary strains, such as wheat with barley, to achieve desired traits. Among the possibilities are plants that can be grown with little or no fertilizer, do not require the application of pesticides, and are resistant to salinity, herbicides, drought, and extreme temperatures. Within the next century it may be possible to develop single plants that "have edible spinach-like leaves, high protein seeds similar to beans, highly nutritious tubers like potatoes, stems that yield useful fiber, and roots that have nitrogen-fixing qualities."[46] Such plants would not only make it possible to use agricultural land and water much more efficiently but also to conserve resources used in the production of fertilizers and avoid pollution resulting from the application of pesticides. Comparable breakthroughs may also be possible in combining the genes of unrelated species of animals that could revolutionize the production of livestock. The recent successful transfer of growth hormones from rats to mice suggest possibilities for developing larger, faster-growing species of cattle.

Outer space is another frontier that high technology may bring into the human domain. The spectacular accomplishments of the United States and Soviet Union in the exploration and use of outer space over the past quarter century has led Princeton physicist

Gerard K. O'Neill to envision limitless possibilities in space for alleviating both the population pressures and resource scarcities on the planet earth. To exploit these possibilities he has designed "space colonies," which would be populated by human beings who would use lunar materials to construct gigantic solar collectors from which energy could be beamed to earth via microwaves. This energy from outer space could be transformed into ordinary electricity, thereby satisfying a sizable proportion of humanity's energy needs with a minimum of environmental damage. In the more distant future it may be possible to mine asteroids and some of the satellites of other planets.[47]

O'Neill's first colonies would involve a sphere one mile in circumference, large enough to accommodate a population of ten thousand people. Later, larger colonies could be built with a circumference of as much as forty miles. Rays from the sun would provide all the energy required. Food would be grown in the space vehicle under ideal growing conditions made possible by the ability to manipulate climate and periods of light and darkness and by the absence of agricultural pests and parasites. The colonies would be equipped with an earthlike environment with air, water, trees, flowers, grass as well as gravity and a suitable climate that would attract human inhabitants. O'Neill predicts that a substantial proportion of the human population will be living on space colonies within the next century.[48] Future generations may even be born, live, and die in outer space.

Can high technology bail humanity out of its ecological predicament? For an unrepentant technological optimist like O'Neill, the answer is unequivocally yes. Others of us are less inclined to put faith in what are referred to pejoratively as "technological fixes." Not only is it possible that the new technologies may prove to be elusive, or to require immense investments of capital that is simply not available, but they could have profound and unanticipated impacts on man and his environment, not all of which would be desirable. High technology solutions to global environmental problems represent a gamble in which the stakes would be very high.

Summary and Conclusions

The six designs discussed in the section above offer a diverse group of strategies for addressing the agenda of global environmental problems. Yet they should not be looked upon as mutually exclusive alternatives, for several of them might be merged into a more comprehensive course of action. For example, both Heilbroner and Ophuls view stronger, centralized political authority as being necessary to establish an economic system that incorporates the principles of steady-state economics. Alternatively, the "small is beautiful" theorists would establish steady-state economies through a decentralized political system in which smaller communities would be in control of most of their own affairs. The authoritarian alternative is not necessarily incompatible with the "small is beautiful" approach; strong political authority may be necessary to counteract trends toward centralized economic power and to increase the flow of people to smaller, self-reliant communities. The scenarios most difficult to combine would be steady-state economics and high technology, especially given the confidence of the technological optimists that we can transcend those physical limits that the steady-state economists take as their basic premise. Even here, it may be possible to pursue some of the most promising high technology solutions to ecological problems while basing them on steady-state principles.

It has not been possible in the brief discussion to provide a detailed look at the six types of designs. Nor has it been possible to assess how successful they would be in addressing the environmental components of the global problematique, or to determine whether there is any reasonable possibility that they could be implemented, or to estimate what undesirable and unintended consequences might result that would counteract their benefits. Skeptics of the international regime approach might argue, for example, that international policy processes are too slow to keep up with the rapidly growing agenda of problems. In regard to global equity designs, questions arise as to whether the industrialized world can ever be persuaded that its interests lie in making what appear to be significant concessions to assist the Third World in achieving its development objectives. Research on recombinant DNA raises fears that noxious organisms will be created and reproduce themselves prolifically in an ecosystem that has no natural defenses against them. These are but a few of the many questions that should be asked in critically evaluating each of the designs. But before we dismiss any of these scenarios, we should also scrutinize the existing global order and ask whether we are willing to settle for the type of future that will follow if significant changes are not made.

In this chapter a distinction has been drawn between two types of thought and research about the future: forecasts projecting what kind of future is believed to be most likely, and designs advocating the most desirable among the futures that are possible (but not necessarily probable). The forecasts of global ecological futures range from the pessimistic doomsday scenarios of the Club of Rome reports, which warn that humanity's growth trends will run up against the limits of the planet's natural endowments in the not-too-distant future; to the unbridled optimism of high technology, which argues that there are no insurmountable physical limits to growth. The designs range from a strengthening of UNEP's programs and the further development of international regimes, to science fiction scenarios of space colonization. Some designs would have us live the frugal life-style of a steady-state economic system in small, relatively autonomous communities or neighborhoods. Others would subject us to stronger centralized controls at both the national and international levels. None of these approaches is certain. And none allows for major unanticipated events. But all share the common focus of concern for the future state of our planet.

6 The Environmental Issue Revisited

In this text the authors have described the evolution and nature of contemporary environmental issues affecting the international community, and the current policies and broad strategies for dealing with these problems. The changing emphasis within existing governmental structures and the creation of a new array of actors, both governmental and nongovernmental, reflect a heightened awareness that human activity has consequences that place the environment at risk for both present and future generations. In this chapter, we once more look at the problem as a whole. We begin with the major theories of the causes of environmental degradation. We then focus on the decade since the 1972 Stockholm Conference, addressing both the state of the global environment and the institutions that have assumed responsibility for dealing with it. Finally, we examine again the debate between the pessimists and the optimists, searching for enough common ground to find hope that although the problems of the global environment are real, they are also manageable without radical changes in the quality of life to which we are accustomed. We do not, of course, expect to find solutions to all (or even any of) the environmental problems that have been raised, but we do expect to provide a perspective that shows promise for the future and stresses the role each of us must play in the creation of that future.

Why Is It Happening?

Throughout this book we have touched on the question of why the problem of the environment has grown so severe, although usually we have studiously avoided identifying a culprit. There are several reasons for this avoidance. As one might expect, there is disagreement about the antecedents of the problem, and this disagreement is exacerbated when one tries to encompass all of the problems appearing under the rubric of the global environment. A second reason for not discussing the causes of these problems is that the emphasis (by both actors and observers) has typically been placed on solutions rather than recrimination, that is, on what future action is to be taken if resolution is to occur. Finally, the true cause in any given case probably lies in some combination of conditions, with each contributing to the outcome.

Let us now turn to the factors that have been suggested as accounting for environmental degradation: 1. population, 2. affluence, 3. technology, 4. capitalism, 5. growth.[1]

With the exception of nuclear war—and one might even make a sound argument in that case—dramatic increases in *population* size have served as a necessary condition for virtually every current environmental problem. Increased population has placed demands on energy, food, and other resources, while at the same time contributing to increased pollution. Most scholars who address environmental issues allude to the rapid growth of population as a major factor, particularly the dramatic increases (and rates of increase) in developing societies. Typical of this view is a recent statement by Lester Brown, informed observer of the world's resource base: "Until the rate of population growth slows markedly, improving the human condition will be difficult."[2]

Not all scholars, however, hold that population is a culprit, even in the rapidly growing areas of the developing world. One such author is Julian Simon who is in the forefront of the minority who question the contention that population growth is at fault. Says Simon: "Population growth has positive economic effects in the long run. . . . (It) is a moral and material triumph."[3] Simon himself recognizes that his view deviates widely from the mainstream, causing him to reveal in the introduction to his nontechnical treatment of the subject that "they [his opponents] were near the point of shutting me up and shutting me down."[4] Although it is clear that the population theory of environmental damage has merit, one must look more closely at the causal links before being completely satisfied about the theory's plausibility.[5]

Those who argue that *affluence* has resulted in a wide range of negative environmental effects point to the correlation between major increases in the gross national product per capita in developed Western countries on the one hand, and high levels of pollution, resource depletion, and the destruction of virgin land on the other. The debate over the role played by affluence is much more pronounced than that for population. Although it is true that the correlation between GNP per capita and environmental degradation is strong, the causal aspects of the link are much less certain. Barry Commoner, a noted commentator on environmental issues, argues that the significant increases in pollution cannot be explained simply by increased consumption.[6] For Commoner, production processes are to blame, so the producers of commodities rather than consumers must bear the major burden of responsibility.

Commoner and others thus point to *technology*, the constantly updated methods of production, as the major cause of negative environmental impact. For them, new technological innovations have made the difference: "detergents, aluminum, plastic and inorganic fertilizers in place of soap powder, wood, steel and manure," disposable packaging in place of returnable bottles, more lead in gasoline, and increased chlorine production leading to increases in mercury consumption. Others disagree with this view, arguing that the environmental benefits of technology outweigh any costs. Robin Attfield finds a common ground between the two positions, suggesting that although there have been undeniable benefits, ecological damage since 1945 owes much of its origin to technology.[7] Presumably, therefore, one may address immediate problems with greater technological clarity (although not necessarily greater ease), than the less discernible long-term impacts that may come from the efforts of technology to solve those problems.

Capitalism has always been pointed to by Marxists as the major culprit for all of society's ills. In this case they are joined by many other thoughtful individuals who add the economic form of society to the technical dimension just discussed. The new technology was introduced because it yielded more profits. Consequently, by-products, no matter how unintended or controversial, have been tolerated, particularly when governments have refused to intervene or have been slow to act. Moreover, the link between capitalism and pressure for growth leads to institutional structures that are incompatible with the strategies of steady-state economies and local self-reliance (small is beautiful) outlined in the previous chapter. In response to these criticisms, supporters of capitalism ask why, if capitalism alone is to blame, are there also severe environmental problems in the planned economies of socialist and communist countries?

The final argument, a composite of the others, suggests that *growth* is at the heart of the matter, particularly growth in five areas—population, food production, industrialization, pollution, and consumption of nonrenewable resources. In this view, the linkages among these five contribute to the severity of the problem. This overarching theory forms the basis for the thesis advanced in the *Limits to Growth* reports. Although that thesis is at first compelling, these reports have come under a great deal of criticism and much research remains to be done before the comprehensive, yet somewhat simplistic, growth theory can be embraced.

Thus, we are left with an absence of certainty with respect to the overriding causes of global environmental problems. That uncertainty should not be surprising, nor should it deter salient actors from combating the issues. Indeed, it may simply mean that each environmental issue—whether pollution, resource depletion, decline in genetic diversity, or whatever—has its own discrete set of causes whenever and wherever the problem emerges. For each issue, therefore, there are unique conditions that need to be understood and addressed if the problem is to be solved.

The Last Decade

Throughout this book we have enumerated a growing list of environmental issues. At the same time we have suggested that awareness on the part of both scientists and observers has grown, leading in the post–World War II era to a series of conferences to heighten concern and to propose an agenda of action. The Stockholm Conference of 1972 has been cited as the highwater mark of international environmental concern. This conference—despite inflammatory rhetoric, self-interested distortion of issues, a tendency to compromise to the point of inaction, and uncertainty regarding the extent to which governments would be willing and able to honor their commitments—was nonetheless a milestone. Awareness was stimulated, grievance and tensions were aired, and sufficient cooperation was achieved at least to recognize the need for collaborative action, research, and institutionalization.[8]

One major activity emanating from this conference has been an annual assessment of the environment and, in 1982, a review of the last decade. This latter report, *The World Environment 1972–1982*, produced the following sober analysis.[9]

Atmosphere

The UNEP study drew several conclusions about atmospheric conditions. The first major finding was that carbon dioxide concentrations are slowly and steadily growing, primarily as a consequence of the increasing use of fossil fuels and the clearing of forests. Although it is known that increased carbon dioxide (CO_2) concentrations can affect weather by altering temperature and precipitation patterns, the implications of the effect are still not well understood.

The UNEP study also determined that acid rain is a real phenomenon produced by the burning of fossil fuels, and transported great distances by wind and weather. On the other hand, earlier alarms about the depletion of the ozone layer as a result of supersonic transport and increased chlorofluorocarbons in the stratosphere continue to be reevaluated. On this issue the experts still disagree, and a final conclusion awaits the development of more sophisticated research techniques. There is no indecision, however, about smog (caused by photochemical oxidants). In those cities with effective controls, smog has decreased, but it has increased where controls are absent or ineffective, or where there is a significant increase in automobile use. The same is said about sulfur dioxide and suspended particulate matter concentrations. There also appears to be an increase in stratospheric particulates.

The question of the existence or likelihood in the near future of long-term climatic change remains a debatable one. Not enough is known about the basic processes of carbon, sulfur, and associated elements to predict how climate would be affected. Although many look for a general warming trend, it could not be observed in 1982, according to UNEP. Nor could one observe any marked changes in the variability of weather, local extremes notwithstanding.

The UNEP study also evaluated action taken during the decade to deal with the main atmospheric challenges: improved weather and climate prediction, coping with natural hazards such as floods and droughts, air pollution control, acid rain, CO_2 climate warming, and stratosphere ozone depletion. Most international organizations already identified as having an interest in the atmosphere are becoming increasingly active: WHO, UNEP, ICSU, regional groups such as the UN Economic Commission for Europe (ECE), the Council for Mutual Economic Assistance (CMEA), and the Organization for Economic Cooperation and Development (OECD). These organizations engage in a number of cooperative ventures that are producing better long-term forecasts for weather, increased awareness of the dimensions of the problems of coping with natural hazards, major advances in understanding the technology of air pollution control, first steps toward recognition of the acid rain issue, assessments of ozone layer depletion, and a renewed research effort focusing on the existence and consequences of CO_2 in the atmosphere. With respect to the future, UNEP foresees three major challenges, two relating to forecasting and monitoring, and the third to energy alternatives. Long-range weather and climatic forecasts of more than five days are needed, as are integrated monitoring stations on land and at sea. Finally, critical appraisal of the climatic implications of energy policies, particularly with respect to energy alternatives remains a high priority.

Marine Environment

In recent years major advances have been made in the scientific understanding of the physical and chemical properties of the oceans, and of the circulation of their waters. The situation with respect to the monitoring of pollution and marine ecosystems is different, however, because only a few localities are the object of investigation. Most ocean pollution is due to the dumping of sewage, agricultural chemicals, oil, and metals. Pollution (particularly iron, manganese, copper, zinc, lead, tin, and antimony) arrives in the ocean via rivers and through atmospheric deposition for certain metals and synthetic chemicals. Extensive exploitation of the mineral resources of the seabed within the continental shelf has occurred during the past decade, while coastal zone developments and oil pollution have had a major impact on the shoreline. Although there is a decline in concentrations of both PCBs and DDT in the coastal waters of the developed West (except Japan), both are still widely found. No fish stocks have been reduced through chemical contamination but overfishing has depleted both the Peruvian anchovy and the North Atlantic herring, and other stock have suffered dramatically. However, there are now increases in the levels of some fish stock, and as a consequence of the concern over mammals throughout the 1970s, whale catches have been greatly reduced. That decade produced a substantial number of global and regional conferences designed to combat marine environmental issues. One should not be misled into concluding that pollution has been confined to coastal areas; rather those areas simply represent the primary focus of research. Much more data are needed about the open oceans before major trends can be established and broad generalizations made.

Inland Waters

Inland waters constitute less than .01 percent of global waters. However, these waters have three major uses: household supplies, industry, and agriculture. Of these, irrigation for agricultural purposes represents the largest share, and the total amount of irrigational use rose throughout the decade. Although industry is increasing its overall usage, that increase has been balanced by more efficient utilization. International activity with regard to inland waters has focused on two areas. First, attention has been directed to seasonal and yearly fluctuations in the amount of available water and the man-made changes in the physical and biological character of this water. However, advances have been made primarily in the capability to assess the problem rather than to resolve it. Second, the international community has addressed the question of the large number of people with no access to safe, clean water or to sanitary services. In rural areas, the percentage of people with safe water and with sanitary facilities has risen to levels of 29 percent (from 14 percent in 1970) and 13 percent, respectively. City dwellers also improved their access to safe water during the decade (from 67 to 75 percent), but access to sanitary facilities for them dropped from 71 to 53 percent. Inland water bodies have suffered pollution and overenrichment (that is, too many nutrients) from industrial discharges, drainage from chemicals and wastes, and the consequences from acid rain. In addition, underground aquifers (the pool of water flow and drainage that supplies the major regions of the world) have sustained losses.

There have been many positive aspects as well. Planning has improved, flood forecasting has become more accurate, some major rivers have become less polluted, irrigation practices have been extended, and a vast number of national water development plans have been put into effect.

Mineral Resources

Most of the world's mineral resources are nonrenewable and are found in the earth's crust, the lithosphere. During the decade since Stockholm, scientists have advanced the definition of resources and reserves, and their classifications, concluding that many resource estimates were tentative and influenced strongly by investment factors. Debate has continued over the consequences of increased mineral consumption, including problems of depletion.

Almost all major nonmetallic minerals are produced at ever-expanding annual rates, but metal production has experienced only a moderate and fluctuating rate of increase. Table 6.1 lists the reserves and resources of selected minerals.

The adverse environmental effects of extraction, treatment, and transport of both metallic and nonmetallic minerals have diminished, particularly in dust control, reclamation of open pits, reduction of acid drainage, and treatment of tailings and liquid waste. Recycling and substitution of mineral raw materials have expanded, but these processes are no longer viewed as the panacea once envisaged. Security of supply continues to be a problem for many industrialized countries. Table 6.2 displays the imports of strategic minerals to the United States in 1980, showing a strong dependence on foreign suppliers.

Finally, earthquakes, volcanoes, and landslides continue to claim much damage to human beings and property. The Tangshang earthquake of 1976 in China killed almost a quarter of a million people and injured approximately the same number. Efforts to limit losses to geologic hazards have increased throughout the decade, although the first step has barely been taken to effect meaningful results.

Living Resources

The earth's natural living resources (technically termed the "terrestrial biota") include the populations of wild plants and animals, their major support systems (termed "biomes"), and their interactions with the nonliving aspects of the environment. One fact has become clear since Stockholm. Where development has either been hampered by poverty or driven by great urgency, whole species of plants and animals have disappeared. Tropical rainforests have been destroyed throughout the decade at about an average of 11 million hectares per year. Desertification is also occurring in tropical deciduous forests at an alarming rate. Forests in temperate zones are typically well managed and are thought to be far less affected by man, but now acid rain appears to be a problem of growing and potentially catastrophic proportions. In the Arctic tundra, man has taken his toll on the environment in his search for oil and gas. Areas of special sensitivity, including islands, mountains, and wetlands are under increasing attack from a number of sources. Islands suffer from the introduction of exogenous species, phosphate mining, and tourism. Deforestation, overgrazing, destructive cultivation, fire, road building, and other factors increase erosion in moun-

Table 6.1 Reserves and resources of selected metals and minerals, 1980

Metal	World Reserve Base (unit)	Estimated World Resources
Antimony	4,716,400 (tons)	Main identified world resources estimated at 5,079,200 tons.
Asbestos	87,800,000 (tons)	90 million tons of identified resources and hypothetical resources of 45 million tons.
Bauxite	22,800,000 (thousand dry tons)	World resources (reserves plus sub-economic and undiscovered deposits) estimated at 40-50 billion tons.
Cadmium	680,000 (tons, metal)	World resources estimated at about 9 million tons.
Chromium	3,355,900 (thousand tons)	World resources total about 36 billion tons of shipping-grade chromite.
Cobalt	3,083,800 (tons, metal)	Identified world resources at about 6 million tons; large hypothetical and speculative resources in manganese nodules on the seabed in lateritic iron-nickel deposits.
Copper	493,000,000 (tons, metal)	Total land-based resources including hypothetical and speculative deposits, estimated to contain 1,627 million tons of copper. Additional 690 million tons estimated in deep-sea nodule resources.
Fluorspar	548,735,000 (tons)	Identified world resources approximately 100 million tons of contained fluorine. World resources of fluorine from phosphate rock estimated at 400 million tons.
Gold	31.4 (million kg of metal)	World resources estimated at 62.2 million kilograms, of which 15-20 per cent are by-product resources.

Table 6.1 (continued)

Metal	World Reserve Base (unit)	Estimated World Resources
Iron ore	105,000 (million tons of usable ore, including by-product ore)	World resources estimated to exceed 800 billion tons of crude ore, containing more than 235,820 million tons of iron.
Lead	157,000,000 (tons, lead)	Total, identified, sub-economic world lead resources estimated at about 1.4 billion tons.
Magnesium metal/ compounds	2,607,625 (thousand tons of magnesite)	Virtually unlimited and globally widespread.
Manganese	4,897,800 (thousand tons gross weight)	Identified land-based resources very large but irregularly distributed. Very extensive deep-sea resources in the form of manganese oxide deposits.
Molybdenum	9,843,000 (thousand kilograms metal)	Identified world resources of about 21 thousand million kilograms.
Nickel	54,238,600 (tons, metal)	Identified world resources in deposits averaging 1 per cent nickel or greater are 143 million tons of nickel. Large lower grade deposits and deep-sea resources in manganese nodules.
Phosphate rock	133,000,000 (thousand tons)	World resources in the thousand millions of tons range.
Platinum-group metals (platinum, palladium, iridium, osmium, rhodium, ruthenium)	36,698 (thousand kilograms)	World resources estimated at about 99.5 million kilograms.
Potash	9,100,000 (thousand tons, K_2O equivalent)	Estimated world resources at about 140 billion tons.
Silver	253 (million kg metal)	Resources about double reserves.

Table 6.1 (continued)

Metal	World Reserve Base (unit)	Estimated World Resources
Tin	10,000,000 (tons, metal)	Not available.
Tungsten	2,585,520 (thousand kilograms tungsten content)	Not available.
Vanadium	15,785,280 (thousand kilograms contained vanadium)	World resources exceed 54 billion kilograms.
Zinc	240,000,000 (tons, metal)	Identified world resources estimated at about 1.8 billion tons.

Source: *Mineral Commodity Summaries* (Washington: US Bureau of Mines, 1981); Quoted in Martin W. Holdgate, Mohammed Kassas, and Gilbert F. White, eds., *The World Environment 1972–1982: A Report by the United Nations Environment Programme* (Dublin: Tycooly International, 1982), pp. 180–81.

tain areas. And wetlands are damaged by flooding from a number of sources.

But there is a positive side to the picture as well. Some fourteen multilateral agreements affecting the terrestrial biota were concluded during the 1970s, and other international initiatives are in evidence in a number of other areas as well. Three main strategies are being undertaken with renewed enthusiasm to handle genetic resources: "protected areas for blanket protection of ecosystems and species; genetic conservation for selected economic species—crop plants, livestock and forest trees; and concentration on threatened species of plant and animal."[10] Steps such as these are important if progress is to be made.

Cropland, Rangeland, and Forests

Food production has increased since Stockholm, in part because of more land being brought into production, primarily as a consequence of irrigation. Rangeland has been improved in some areas, but desertification has occurred elsewhere, as has the transfer of land to other uses. In the developed world during the 1970s more than 30,000 square kilometers (km^2) became settlements and roads, and approximately the same amount was lost in the developing world. Desertification resulted in some 60,000 km^2 of land destroyed

Table 6.2 Imports of strategic materials, United States, 1980

Material	Imports as a percentage of apparent consumption	Major sources of imports 1976–1979	Uses
Antimony	53%	China, Mexico, Bolivia	Batteries, solders, flame-proofing
Bauxite	94%	Jamaica, Guinea, Surinam	Packaging, construction, transportation vehicles
Cadmium	62%	Canada, Australia, Mexico	Electroplating, electrodes
Chromium	91%	South Africa, Philippines, USSR	Stainless steel
Cobalt	93%	Zaire, Belgium, Luxembourg, Zambia	Jet engines
Columbium	100%	Brazil, Canada, Thailand	Alloy steels, cutting tools
Manganese	97%	Gabon, Brazil, Australia, South Africa, France	Steel and iron making
Platinum group metals	87%	South Africa, USSR, United Kingdom	Coatings, catalytic converters
Tantalite	97%	Thailand, Canada, Australia	Electronics
Titanium	100%	Australia, India	Titanium sponge products, welding rod coatings
Tungsten	54%	Canada, Bolivia, South Korea	Machine tools, alloys

Source: *Mineral Commodity Summaries* (Washington: US Bureau of Mines, 1981), quoted in Martin W. Holdgate, Mohammed Kassas, and Gilbert F. White, eds., *The World Environment 1972-1982: A Report by the United Nations Environment Programme* (Dublin: Tycooly International, 1982), pp. 197.

or impaired annually. Soil degradation continues to be a major problem in many parts of the globe.

There is also continuing concern over the adverse effects of agricultural chemicals. Nitrogenous fertilizer production almost doubled during the 1970s, and nitrates from fertilizer residues polluted waters. Chemical pesticide usage has increased with harmful effects on living beings. Environmental testing of new pesticides means a longer lead time between discovery and introduction into the market-place, while at the same time the pests themselves have developed stronger resistance to traditional pesticide types. In reviewing the decade following Stockholm, there was basic agreement that more action is needed "to stop desertification, to conserve soil, to manage forests and water wisely, and to use appropriate methods of cultivation."[11]

Population

The above categories of environmental concern address primarily natural processes, but human activities have also had a major impact on those processes. The two final issues, population and war, are clearly generated by humans. Population has been an issue of mounting concern since the 1950s. Three population processes are continuously at work: fertility, mortality, and migration. Population problems occur because these three dynamic processes interact to produce change in population structures—size, age composition, and distribution—resulting in some related social, economic, or political problems. Population thus becomes linked to environmental issues: not enough arable land, too much pollution, too rapid depletion of nonrenewable resources, not enough housing, too few jobs, and so forth.

During the 1970s the world's population increased 700 million to over 4.4 billion. This rate of increase, 1.72 percent, meant that one million additional people were added to the planet every five days. In the developed world, the rate of growth actually declined from .84 percent at the beginning of the decade to .71 percent at the end. The developing world, however, had a much higher growth rate, 2.08 percent in 1980, although it too declined from 2.32 percent in the first half of the decade. The UN has projected that the rate of growth in the developing world will continue to drop to a rate of 1.84 percent by the end of the century. The bad news about this prediction is that the population in that part of the world will still double every thirty-eight years. While fertility in the developing world is declining, mortality is dropping even faster; those born during the baby boom of the developing world, the mid-1960s, have not yet reached their life expectancy. Moreover, since infant mortality has declined significantly in the Third World—although it still remains much higher than in the developed world—because of the diffusion of health care from the West, the percentage of the population under the age of twenty was substantial. Table 6.3 ranks the ten youngest countries. As you can observe, the median age

Table 6.3 Ten youngest countries

Rank	Country	Population median age	Percent of population under age 15	Percent of population aged 65 or more
1	Botswana	14.2	49.9	2.6
2	Kenya	14.3	51.8	2.6
3	Liberia	15.8	48.4	2.6
4	Malawi	15.9	48.2	2.6
5	Syria	16.0	48.0	2.9
6	Nigeria	16.1	47.8	2.5
7	Zimbabwe	16.1	47.7	2.7
8	Algeria	16.3	47.1	3.3
9	Ghana	16.4	47.1	2.3
10	Libya	16.7	46.4	2.8

Source: Leon F. Bouvier, "Planet Earth 1984–2034: A Demographic Vision," *Population Bulletin* 39, 1 (February 1984): 15.

(half the population is above and below this level) is 14.2 for the youngest (Botswana) and even as low as 16.7 for the country in tenth position (Libya). The presence of a young population is a primary problem for the developing world because limited public funds and a low level of development mean that the society cannot provide adequately for these individuals until the day when they can begin to make an economic contribution to society.

In the last decade, major advances have been made by Third World governments in introducing either population planning—whereby the government uses incentives and disincentives to convince citizens to reproduce at government-prescribed levels—or family planning, a less interventionist policy based on programs to educate citizens and provide the means for them to achieve their own population goal. There are 132 developing countries with populations over 100,000. Thirty-five have official population planning programs, and 31 have family planning policies. Thus slightly more than half of the Third World countries have official policies, but these nations contain 91 percent of the population that reside in the poor countries (77 percent in population planning countries and 14 percent in family planning). It remains to be seen how effective these programs will be because an overwhelming majority of women in the Third World desire a significantly higher number of children than the 2.1 or 2.2 replacement rate for couples. Recent evidence from China suggests that even its presumably successful program still results in a large percentage of women exceeding the government-prescribed level.

The third population process, migration (both internal and international), has also undergone changes in the last decade. Indeed, migration may become the major population problem of the twenty-first century. Within the developing world, movement to the urban areas is increasing dramatically. Seven Third World urban areas now make the top ten list of most populous cities (Mexico City, Shanghai, São Paulo, Beijing, Greater Buenos Aires, Rio de Janeiro, and Seoul).[12] It is estimated that there will be an additional three hundred cities in this part of the world with over one million people by the end of the century.[13]

Movement of people, both workers and refugees, between nations (international migration) continues to grow: legal migration, refugee flights, and illegal "revolving door" migration (back and forth many times). The migration of labor (both semiskilled and professional), primarily from the South to the North, slowed down during the 1970s. This kind of migration will probably continue to slow down in the short run as governmental policy becomes more restrictive. It is the movement of refugees that contributes to the high migration figures. In 1982 there were over 14 million refugees in the world, many scattered throughout Africa. The focus of migration has thus shifted from Southeast Asia, although the latter still has some six million refugees.[14] Although the United Nations has been receptive to the problem of migration, the number of refugees and transients continued to grow without significant impact from policy intervention.

War

According to the UNEP study both conventional and nuclear wars represent potential environmental problems. These fall into three categories: environmental consequences of current and past wars (hazards from unexploded weapons, physical and biological effects of damage to soil and landscape, human suffering); environmental consequences of preparations for war (diversion of resources from environmental development, impact of the armaments industry, weapons testing, military operations, proliferation of nuclear technology); and environmental hazards of future war including conventional, nuclear, chemical, and biological warfare.

With respect to the first category—*environmental consequences of current and past wars*—unexploded munitions still remain a major problem during the decade. Some 150,000 to 300,000 remain in Vietnam, and one government report states that it continues to clear 300,000 to 400,000 bombs per year left from World War II. Concrete roadblocks, gun emplacements, abandoned airfields, and other installations constitute a problem as well. In Vietnam, chemical herbicides were responsible for the destruction of some 1,500 km^2 of mangrove forest and damage to a similar amount. One study suggested that there is no guarantee that original ecosystems will ever be restored.[15] The massive disruption of vegetation, followed by widespread soil erosion, severe changes to the water support system and to the pattern of returning plants suggest that restoration will be difficult if not impossible. But even more tragic than such damage to the physical environment was the disruption of social life caused by war. Close to 30 million civilians were killed in World War II, for example, while the Kampuchean insurrections of 1975–77 resulted in over a million civilian deaths.

Preparation for war can also affect environment issues, although less directly than war itself. In real terms, the rate of increase in global military expenditures was less during the 1970s than during the previous decade. However, the fact remains that the annual level of global military expenditures now totals more than $600 billion. Using this indicator to represent the problem of preparation for war, the scope has expanded in two ways. Vertically, there have been increases in total expenditures. Horizontally, many more countries are now engaged in high levels of spending. Within the developed world, there have been continual increases in military spending. In addition, developing countries have also increased their spending, largely in the form of arms purchases from the developed world. Unfortunately, those developing countries that could least afford to do so were diverting funds away from social programs to purchase arms. The developing world spent close to 20 percent of the world military budget, double the percentage of twenty years ago. This increase obviously meant dramatic growth in the international movement of armaments. And as a consequence, the oceans, stratosphere, and space became increasingly militarized.

The production of weapons as an industrial process contributes to environmental pollution. If one considers that military expenditures are between 5 and 10 percent of the GNP, then that proportion of pollution caused by industry can be attributed to arms production. The problem is that all other things are not equal, because the defense industry handles particularly hazardous materials. Between 1970 and 1980 there were 469 nuclear explosions, mostly tests.

Forty-one were atmospheric tests conducted by countries that did not sign the Limited Test Ban Treaty of 1963. This treaty improved the situation in atmospheric pollution as measured, for example, by generally falling levels of radio-nuclides in rain and in milk (caused by the ingestion of grass by cattle that have absorbed rain-borne fallout). Other kinds of advanced weapons testing continue, although much less is known about their environmental impacts.

With respect to the *hazards of future war*, it is not only the possibility of nuclear conflict that is feared by environmentalists. The destructive level of conventional weapons grew steadily throughout the 1970s; aircraft carrying high explosive cluster bombs or grenade clusters could deal a blow comparable to a tactical nuclear guided missile with a one kiloton warhead. Chemical and biological warfare deliberately pollutes the environment by the release of toxic chemicals or harmful microorganisms. At the end of the decade, chemical weapons, which had once been banned, were again part of military planning. Although prohibited by international agreements, research and development on chemical weapons continue, and there is mounting evidence that such weapons have been used in several regional conflicts in the 1980s. In addition, military planners have also considered environmental modification techniques, a direct assault on the environment as a method of disrupting the enemy. Table 6.4 reveals the feasible techniques available at the close of the decade. Nuclear war, however, remains the greatest environmental hazard. The effects of even a single nuclear explosion are well documented. Table 6.5 shows the immediate impact of a one-megaton explosion, and table 6.6 reveals the twenty-four hour effects of a ground burst (not the most lethal) of two larger bombs.

The most comprehensive study ever conducted on the climatic and biological effects of large-scale nuclear war was recently completed by a team of scientists headed by Carl Sagan, astronomer and space scientist. Their major finding is that "the long-term consequences of a nuclear war could constitute a global climatic catastrophe."[16] Nuclear arsenals of strategic, theater, and tactical weapons number close to 50,000 for the United States and Soviet Union, and the rest of the world may have several hundred. The study team identified four long-range environmental effects that occur after the nuclear war is over: "obscuring smoke in the troposphere, obscuring dust in the stratosphere, the fallout of radioactive debris, and the partial destruction of the ozone layer."[17] Severe and long-term low temperatures (at least four months) would result, even

Table 6.4 Hostile environmental modification techniques feasible in 1980*

Basic natural systems	Type of effect	Military significance	Remarks
Atmosphere	Dispersion of cloud cover or fog	Ensures visibility in combat areas, airfields and naval bases	Effective in limited areas as a tactical device
	Artificial creation of fog or clouds	Impedes surveillance by the enemy, and protects against light radiation from nuclear explosions	Effective in limited areas, under certain weather conditions as a tactical device
	Artificial creation of hail, snow or rain	Damage to communications equipment and certain types of military equipment	Effective at certain altitudes in limited areas as a tactical device
Inland Waters	Changes in local water balance (destruction of dykes and irrigation works)	Impedes combat operations and logistical support	Effective on a tactical scale
	Action to affect the physical properties of water resources (pollution, infection)	Impedes combat and supply operations, disrupts operations in logistic areas	
Continental Ecosystems	Action to affect permafrost areas	Destroys road networks and airfields, damages water systems	Possible for certain areas
	Stimulation of avalanches and landslides	Destroys road networks and impedes military operations	Possible only in limited areas as a tactical device
	Destruction of vegetation or soil cover	Impedes activities of enemy forces, disrupts agricultural activities	Effective as a tactical device

*Often under limited conditions only
Source: Martin W. Holdgate, Mohammed Kassas, and Gilbert F. White, eds., *The World Environment 1972–1982: A Report by the United Nations Environment Programme* (Dublin: Tycooly International, 1982), p. 609.

Table 6.5 The environmental impact of a one-megaton nuclear explosion

Distance (km)	Effect
2.4	Reinforced concrete multistorey buildings destroyed. Most people killed instantly.
4.6	Concrete buildings destroyed. Spontaneous ignition of clothing. Third-degree flash burns to exposed skin.
6.7	Brick and wood frame houses destroyed. Spontaneous ignition of clothing. Third-degree flash burns to exposed skin.
7.8	Spontaneous ignition of clothing and other combustion materials. Third-degree flash burns to exposed skin.
9.9	Third-degree flash burns to exposed skin.
13.6	Moderate damage to brick and wood frame houses.

Source: *The Effects of Nuclear War* (Washington: U.S. Arms Control and Disarmament Agency, 1979), quoted in Martin W. Holdgate, Mohammed Kassas, and Gilbert F. White, eds., *The World Environment 1972–1982: A Report by the United Nations Environment Programme* (Dublin: Tycooly International, 1982), p. 607.

from a comparatively small nuclear war, followed by an extended period of increased ultraviolet light on the earth's surface. Widespread fires and subsequent runoff of topsoil would result, starvation would be rampant, and radiation sickness would be evident everywhere, not just in the areas of the bomb blasts themselves. The study concluded:

> Species extinction could be expected for most tropical plants and animals, and for most terrestrial vertebrates of north temperate regions, a large number of plants, and numerous freshwater and some marine organisms. . . . Whether any people would be able to persist for long in the face of highly modified biological communities; novel climates; high levels of radiation; shattered agricultural, social, and economic systems; extraordinary psychological stresses; and a host of other difficulties is open to question. It is clear that the ecosystem effects alone resulting from a large-scale thermonuclear war could be enough to destroy the current civilization in at least the Northern Hemisphere. Coupled with the direct casualties of perhaps two billion people, the combined intermediate and long-term effects of nuclear war suggest that eventually there might be no human survivors in the Northern Hemisphere.

Furthermore, the scenario prescribed here is by no means the most severe that could be imagined with present world nuclear

Table 6.6 Some effects of ground-burst nuclear weapons within 24 hours of detonation

Type of damage	Area over which damage may occur (ha) 20-kiloton atomic bomb	10-megaton hydrogen bomb
Craterization by blast wave[a]	1	57
Vertebrates killed by blast wave[b]	24	1 540
All vegetation killed by nuclear radiation[c]	43	12 100
Trees killed by nuclear radiation[d]	148	63 800
Trees blown down by blast wave[e]	362	52 500
Vertebrates killed by nuclear radiation[f]	674	177 000
Dry vegetation ignited by thermal radiation[g]	749	117 000
Vertebrates killed by thermal radiation[h]	1 000	150 000

[a] Refers to dry soil. A sub-surface burst could craterize four times as large an area as a surface burst. Nuclear warheads exploding above the surface would produce no craters at all if the burst were sufficiently high, but the nuclear and thermal radiation effects would then extend over larger areas.

[b] Refers to areas over which the transient over-pressure, would be likely to exceed 345 kilopascal; this being the over-pressure for about 50 per cent lethality among large mammals, including man. These figures are, in fact, augmented by reflected over-pressures which (depending upon height of burst, terrain, etc.) can more than double the total (so-called Mach front) over-pressures experienced at any distance.

[c] Refers to areas over which the early radiation dose would be likely to exceed 70 kilorad.

[d] Refers to areas over which the early radiation dose would be likely to exceed 10 kilorad.

[e] Refers to areas over which the transient wind velocity at the shock front, ignoring the Mach front, would be likely to exceed about 60 metres per second. Such a wind would be likely to blow down about 90 per cent of the trees in an average coniferous forest or a deciduous forest in leaf.

[f] Refers to areas over which the early radiation dose would be likely to exceed 2 kilorad.

[g] Refers to areas over which the incident thermal radiation would be likely to exceed 500 kilojoules per square metre for the atom bomb or 1,000 kilojoules per square metre for the hydrogen bomb if the weapons were detonated on a clear day having a visibility of 80 km.

[h] Refers to areas over which, on a clear day of 80 km visibility, the incident thermal radiation would be likely to exceed that which would have a 50 per cent lethality for exposed pigs (380 kilojoules per square metre for the atom bomb and 750 kilojoules per square metre for the hydrogen bomb).

Source: J. P. Robinson, *The Effects of Weapons on Ecosystems*, United Nations Environment Programme Studies, vol. 1 (New York: Oxford University Press, 1979), quoted in Martin W. Holdgate, Mohammed Kassas, and Gilbert F. White, eds., *The World Environment 1972-1982: A Report by the United Nations Environment Programme* (Dublin: Tycooly International, 1982), p. 608.

> arsenals and those contemplated for the near future. In almost any realistic case involving nuclear exchanges between the superpowers, global environmental changes sufficient to cause an extinction event equal to or more severe than that at the close of the Cretaceous when the dinosaurs and many other species died out are likely. In that event, the possibility of the extinction of Homo sapiens cannot be excluded.[18]

The international community has responded to the threat of war-related destruction by reaching agreement on a number of arms control treaties during the decade, while keeping some previous ones in place. Table 6.7 lists agreements in force at the end of the decade. Others await ratification or are under negotiation at this time; although progress has been slow on both multilateral and bilateral (U.S. and USSR) arrangements.

UNEP Today

In May 1982, following the report reviewing the last decade, UNEP held a special session in Nairobi, Kenya, to commemorate the tenth anniversary of the Stockholm Conference. The adopted resolution at this special session suggested that "significant progress" had been made but "the Action Plan (of Stockholm) [had] only been partially implemented and the results [could not] be considered satisfactory."[19] As UNEP's Executive Director, Mustafa Tolba stated: "Our room for maneuver has narrowed considerably since 1972. . . . On virtually every front there has been marked deterioration in the quality of everything. . . . [T]he planet's capacity to meet [its] needs is being undermined." He concluded: "In 1982, nations have two choices: to carry on as they are and face by the turn of the century an environmental catastrophe which will witness devastation as complete, as irreversible as any nuclear holocaust, or to begin now in earnest a cooperative effort to use the world's resources rationally and fairly."[20]

Many pointed to a decline in the commitment of resources, particularly from the wealthier countries that were leaders at Stockholm. The world economy did not enjoy a successful decade in the 1970s and this was reflected in donor contributions. The United States was singled out as one of those countries whose annual contribution (10 million dollars) has declined. On the other hand, developing countries whose role at Stockholm was one of suspicion, argued at Nairobi that strong action must be taken to combat

Table 6.7 Principal international arms regulation or disarmament agreements currently in force

Title of agreement	Signed	Entered into force	Number of states party[a]
Protocol for the Prohibition of the Use in War of Asphyxiating, Poisonous or Other Gases, and of Bacteriological Methods of Warfare	1925	1928	96
The Antarctic (de-militarization) Treaty	1959	1961	20
Treaty Banning Nuclear Weapon Tests in the Atmosphere, in Outer Space and Under Water	1963	1963	111
Treaty on Principles Governing the Activities of States in the Exploration and use of Outer Space, including the Moon and Other Celestial Bodies	1967	1967	80
Treaty for the Prohibition of Nuclear Weapons in Latin America	1967		27
Treaty on the Non-Proliferation of Nuclear Weapons	1968	1970	111
Treaty on the Prohibition of the Emplacement of Nuclear Weapons and Other Weapons of Mass Destruction on the Sea Bed and the Ocean Floor and in the Subsoil Thereof	1971	1972	68
Convention on the Prohibition of the Development, Production and Stock-piling of Bacteriological (Biological) and Toxin Weapons and on Their Destruction	1972	1975	87
US-Soviet Treaty on the Limitation of Anti-Ballistic Missile Systems	1972	1972	2
Convention on the Prohibition of Military or Any Other Hostile Use of Environmental Modification Techniques	1977	1978	27

[a] As of 31 December 1979. Source of Table: United Nations. The full texts of the multilateral treaties are to be found in the United Nations (1978 c) and USACDA (1977).

Source: Martin W. Holdgate, Mohammed Kassas, and Gilbert F. White, eds., *The World Environment 1972–1982: A Report by the United Nations Environment Programme* (Dublin: Tycooly International, 1982), p. 614.

environmental problems. But perhaps the central conclusion to be drawn can be found in the words of Gilbert F. White, an environmental scientist: "The decade after Stockholm has shown that environmental improvement can be achieved, that the pace is slow in many areas, and that scientific inquiry can help speed it up."[21] Finally, the special session concluded that UNEP should continue to play a catalytic and coordinating role in monitoring and assessing problems, in establishing appropriate policies and programs, and in developing educational programs.

Conclusion

The analysis of the last decade reveals that problems exist and that many of them are increasing in severity, but assessments of the character and the cause of these problems have also been increasing. The decade has been marked, in addition, by a number of global modeling studies that have attempted to acquire a comprehensive picture of the world. Although these models differ greatly with respect to their purposes, techniques, findings, forecasts, and prescriptions, they have drawn surprisingly similar conclusions about the present and the future of the world as summarized in a recent U.S. Office of Technology Assessment study:

(1) Population and physical capital cannot grow indefinitely on a finite planet without eventually causing widespread hunger and resource scarcities. However, there is no physical or technical reason why basic human needs could not be supplied to all the world's people, now and for the foreseeable future. These needs are not now being met because of unequal distribution of resources and consumption—not overall physical scarcities. The absence of physical limits, however, does not necessarily imply the existence of a practical solution.

(2) The continuation of current trends would result in growing environmental, economic, and political difficulties. As a result, "business as usual" is not a palatable future course. Technological progress is expected (and in fact essential), but no set of purely technical changes tested in the models was sufficient in itself to bring about a completely satisfactory outcome. The models suggest that social, economic, and political changes will also be necessary.

(3) Over the next two or three decades, the world's socioeco-

nomic system will be in a period of transition to a state that will be significantly different from the present. However, the shape of this future state is not predetermined—it is a function of decisions and changes being made now.

(4) Because of the complexity, momentum, and interdependency inherent in the world's physical and socioeconomic systems, the full long-term effects of a given action are almost impossible to predict with precision or certainty. However, actions taken soon are likely to be more effective and less costly than the same set of actions taken later, and cooperative long-term approaches are likely to be more beneficial for all parties than competitive short-term approaches.

(5) Many existing plans and agreements—particularly complex, long-term international development programs—are based on assumptions about the world that are either mutually inconsistent or inconsistent with physical reality.

(6) Pollution and resource availability may or may not be problems on a global scale, but there is general agreement that regional problems of global concern—such as food shortages in South Asia and perhaps Sub-Saharan Africa—are far more likely than a global collapse.[22]

As can be seen, the situation is constantly evolving. It is evident that the future promises more of the same and at a faster rate. That can be either good news or bad news. In the past, the human race has been fortunate because no single catastrophe could cause annihilation. But we are now fast becoming a single complex, interrelated ecosystem, which means that in the future a single major catastrophe could trigger a number of others and the human race could indeed face destruction. Kenneth Boulding terms this condition as living on a cliff rather than in a lifeboat.[23]

But we must face the future with optimism. One writer has summed it up well by suggesting that "human beings will, in their own selfish best interests, see to it that political leaders are not allowed to introduce warfare among the nations of the earth, so that we all survive; and see to it that all residents of the planet share in the benefits of developments in technology and the sciences. Humankind can solve the problems facing planet earth if humankind so desires."[24]

Notes

Preface

1 Martin W. Holdgate, Mohammed Kassas, and Gilbert F. White, eds., *The World Environment 1972–1982: A Report by the United Nations Environment Programme* (Dublin: Tycooly International, 1982), p. 4.
2 Ibid., p. 5.
3 Ibid.

Environment as a Global Issue

1 Garrett Hardin, "The Tragedy of the Commons," *Science* 162 (3 December 1968): 1243–48.
2 Roy A. Rappoport, *Pigs for the Ancestors: Ritual in the Ecology of a New Guinea People* (New Haven: Yale University Press, 1967).
3 Council on Environmental Quality and the Department of State, *The Global 2000 Report to the President, Volume II: The Technical Report* (Washington, D.C.: U.S. Government Printing Office, 1980), p. 281.
4 Ibid., p. 283.
5 Ibid., p. 277.
6 Ibid., p. 279.
7 U.S. Department of Agriculture and the Council on Environmental Quality, *National Agricultural Lands Study: Final Report 1981* (Washington, D.C.: U.S. Government Printing Office, 1981), pp. 35–37.
8 *The Global 2000 Report Volume II*, p. 281.
9 Council on Environmental Quality, *Environmental Quality—1979* (Washington, D.C.: U.S. Government Printing Office, 1980), p. 609.
10 Ibid., pp. 609–10.
11 Erik P. Eckholm, *Disappearing Species: The Social Challenge* (Washington, D.C.: Worldwatch Institute, July 1978), pp. 6–7.
12 *Environmental Quality—1979*, p. 610.
13 Eckholm, *Disappearing Species*, p. 10.
14 Ibid., p. 16.
15 Office of Technology Assessment, U.S. Congress, *The Effects of Nuclear War* (Washington, D.C.: U.S. Government Printing Office, 1979), pp. 35–38.
16 Ibid., pp. 112–14.

17 Donnella H. Meadows et al., *The Limits to Growth* (New York: Universe Books, 1972).

Environmental Actors

1 *Report of the United Nations Conference on the Human Environment, Stockholm, 5-16 June 1972*, UN Document A/Conf. 48/14/Rev. 1 (New York, 1973), contains the texts of the Declaration, the Action Plan, and the Resolution on Institutional and Financial Arrangements adopted by the conference.
2 For an account of NGO involvement in UNCHE (the Stockholm Conference), see Anne Thompson Feraru, "Transnational Political Interests and the Global Environment," *International Organization* 28 (Winter 1974): 31–60.
3 For a summary of the major global conferences of the 1970s, see A. LeRoy Bennett, *International Organizations* (Englewood Cliffs, N.J.: Prentice-Hall, 1980), pp. 310–36.
4 Harlan Cleveland, *The Third Try at World Order* (Princeton, N.J.: Aspen Institute for Humanistic Studies, 1976), p. 78.
5 For a fuller account of UNEP's role in the Desertification Conference, see *The United Nations Environment Program* (Nairobi: UNEP, 1979), pp. 49–51.
6 "UN Conference on New and Renewable Sources of Energy," *U.S. Department of State Bulletin* (January 1982), p. 66.
7 *Declaration Adopted by the UNEP Session of a Special Character—Nairobi, 18 May 1982* (Nairobi, Kenya: UNEP Liaison Office, 1982), p. 2.
8 Ibid., p. 1.
9 Patricia Scharlin, "Nairobi Remembers the Stockholm Conference," *Sierra* (September–October 1982), p. 28.
10 *Declaration*, p. 3.

Environmental Values

1 Marvin S. Soroos, "Trends in the Perception of Ecological Problems in the United Nations General Debates," *Human Ecology* 9, 1 (1981): 32–35.
2 David Ehrenfeld, *The Arrogance of Humanism* (New York: Oxford University Press, 1978), pp. 208–9.
3 Erik P. Eckholm, *The Picture of Health: Environmental Sources of Disease* (New York: W. W. Norton, 1977), pp. 140–46.
4 W. Eugene Smith, *Minamata* (New York: Holt, Rinehart and Winston, 1975).
5 Norman Myers, *The Sinking Ark: A New Look at the Problem of Disappearing Species* (Elmsford, N.Y.: Pergamon Press, 1979), p.48.
6 Ibid., p. 57.
7 Ibid., pp. 60–61.
8 Ibid., pp. 68–79.
9 Martin W. Holdgate, Mohammed Kassas, and Gilbert F. White, eds., *The World Environment 1972–1982: A Report by the United Nations Environment Programme* (Dublin: Tycooly International, 1982), pp. 230–42.
10 Jerry Rifkin, *Entropy: A New World View* (New York: Viking Press, 1980), p. 6.
11 Garrett Hardin, "Living on a Lifeboat," *Psychology Today* 24, 10 (October 1974): 563–65.
12 Garrett Hardin, *The Limits to Altruism: An Ecologist's View of Survival* (Bloomington: Indiana University Press, 1977), p. 98.

13 "World Abortion Trends," *Population* 9 (April 1979): 1.
14 Ruth Leger Sivard, *World Military and Social Expenditures—1983* (New York: Institute for World Order, 1983), p. 6.
15 Roy L. Prosterman, *Surviving to 3000: An Introduction to the Study of Lethal Conflict* (Belmont, Calif.: Wadsworth, 1972), p. 39.
16 Arthur H. Westing, "Indochina: Prototype of Ecocide," in *Air, Water, Earth, Fire: The Impact of the Military on World Environmental Order*, by Michael McClintock et al. (San Francisco: Sierra Club, 1974), pp. 16–23.
17 Quoted in Alfred E. Eckes, Jr., *The United States and the Global Struggle for Minerals* (Austin: University of Texas Press, 1979), p. 59.
18 Ibid., p. 64.
19 Ibid., p. 70.
20 Lester R. Brown, *Redefining National Security*, Worldwatch Paper 14 (Washington, D.C.: Worldwatch Institute, 1977).
21 Lawrence Juda, "International Environmental Concern: Perspectives of and Implications for Developing States," in *The Global Predicament: Ecological Perspectives on World Order*, edited by David W. Orr and Marvin S. Soroos (Chapel Hill: University of North Carolina Press, 1979), pp. 92–95.

Environmental Policies

1 J. A. Gulland, "Open Ocean Resources," in *Fisheries Management*, edited by Robert T. Lackey and Larry A. Neilsen (New York: John Wiley, 1980), p. 349.
2 H. Gary Knight, *Managing the Sea's Living Resources: Legal and Political Aspects of High Seas Fisheries* (Lexington, Mass.: Lexington Books, 1977), p. 46.
3 Ross D. Eckert, *The Enclosure of Ocean Resources: Economics and the Law of the Sea* (Stanford, Calif.: Hoover Institution Press, 1979), p. 17.
4 Lester Brown, *The Twenty-Ninth Day* (New York: W. W. Norton, 1978), p. 17.
5 Millie Payne, "Status of Whales," *The Living Wilderness* 43, 147 (December 1979): 16–17.
6 Geoffrey Murray, "Japan, Angered by World Whaling Ban, Gets Ready to Defy It," *Christian Science Monitor* (27 July 1982), p. 1.
7 Norman Myers, *The Sinking Ark: A New Look at the Problem of Disappearing Species* (Elmsford, N.Y.: Pergamon Press, 1979), p. 31.
8 Lewis Regenstein, *The Politics of Extinction: The Shocking Story of the World's Endangered Species* (New York: Macmillan, 1975), pp. xiii, xix–xx.
9 Paul Ehrlich and Ann Ehrlich, *Extinction: The Causes and Consequences of the Disappearance of Species* (New York: Random House, 1981), p. 115.
10 Regenstein, *Politics of Extinction*, p. xxii.
11 Ehrlich and Ehrlich, *Extinction*, p. 116.
12 Gregory James, "Yemeni Wealth Decimates Rhinos," *New York Times* (9 March 1982), p. E19.
13 Noel Grove, "Wild Cargo: The Business of Smuggling Animals," *National Geographic* 159, 3 (March 1981): 302.
14 Ibid., p. 300.
15 Myers, *Sinking Ark*, p. 82.
16 Erik P. Eckholm, *Down to Earth: Environment and Human Needs* (New York: W. W. Norton, 1982), p. 194.

17 Bayard Webster, "Tanzania Asks Rich Countries to Aid Wildlife," *New York Times* (29 March 1979), p. 16.
18 Michael T. Kaufman, "Preserving Rare Species Is an Ironic Success," *New York Times* (8 March 1981), p. 20.
19 Eckholm, *Down to Earth*, p. 194.
20 Robert Boardman, *International Organization and the Conservation of Nature* (Bloomington: Indiana University Press, 1981), p. 96.
21 *Petroleum in the Marine Environment* (Washington, D.C.: Academy of Sciences, 1975), p. 6.
22 Joseph E. Brown, *Oil Spills: Danger in the Sea* (New York: Dodd Mead, 1978), pp. 54–55.
23 Walter Kiechel III, "Admiralty Case of the Century," *Fortune* 99 (23 April 1979): 78–82.
24 Richard W. Wagner, *Environment and Man*, 3d ed. (New York: W. W. Norton, 1978), p. 418.
25 R. Michael M'Gonigle and Mark W. Zacher, *Pollution, Politics, and International Law: Tankers at Sea* (Berkeley: University of California Press, 1979), p. 16.
26 Group of Experts on the Scientific Aspects of Marine Pollution, *Impact of Oil on the Marine Environment* (Rome: Food and Agriculture Organization, 1977), pp. 75–76.
27 M'Gonigle and Zacher, *Pollution, Politics, and International Law*, p. 39.
28 Ibid., pp. 226–38.
29 Gene E. Likens et al., "Acid Rain," *Scientific American* 241, 4 (October 1979): 42.
30 Jeffrey Knight, "Clear Air III: The Global Dimension," *Sierra Club Bulletin* 66, 3 (May/June 1981): 19–20.
31 "Acid Precipitation," *Mother Earth News* 73 (January/February 1982): 123.
32 Anne LaBastille, "Acid Rain: How Great the Threat?" *National Geographic* 160, 5 (November 1981): 669.
33 Ibid., pp. 670–72.
34 Ibid., p. 657.
35 For the text of the decision, see James Barros and Douglas M. Johnston, *The International Law of Pollution* (New York: Free Press, 1974), pp. 177–95.

The Future of the Environment

1 Marvin S. Soroos, "Exploring Global Ecological Futures," in *The Global Predicament: Ecological Perspectives on World Order*, edited by David W. Orr and Marvin S. Soroos (Chapel Hill: University of North Carolina Press, 1979), p. 41.
2 Ibid., pp. 42–43.
3 Ibid., pp. 44–46.
4 Ibid., pp. 45–46.
5 Ibid., pp. 49–50.
6 Donella Meadows et al., *The Limits to Growth* (New York: Universe Books, 1972).
7 Ibid., p. 23.
8 Ibid., p. 127.
9 Ibid., p. 136.
10 Ibid., p. 138–39.

11 Mihajlo Mesarovic and Edward Pestel, *Mankind at the Turning Point* (New York: E. P. Dutton, 1974), p. 55.
12 Ibid., p. 75.
13 Ibid., p. 55.
14 Council on Environmental Quality and the Department of State, *The Global 2000 Report to the President, Volume I* (Washington, D.C.: U.S. Government Printing Office, 1980), p. 1.
15 Ibid., p. 3.
16 Wilfred Beckerman, *Two Cheers for the Affluent Society: A Spirited Defense of Economic Growth* (New York: St. Martin's Press, 1974), pp. 173–79.
17 Ibid., p. 173.
18 Ibid., pp. 178–89.
19 Julian L. Simon, *The Ultimate Resource* (Princeton, N.J.: Princeton University Press, 1981), p. 7.
20 Ibid., pp. 17–20.
21 Ibid., p. 6.
22 Ibid., p. 7.
23 Herman Kahn, William Brown, and Leon Martel, *The Next 200 Years: A Scenario for America and the World* (New York: William Morrow, 1976), p. 7.
24 Ibid., pp. 7–8.
25 Ibid., p. 10.
26 Katherine Gilman, "Julian Simon's Cracked Crystal Ball," *The Public Interest* 65 (Fall 1981): 74–77.
27 Kenneth D. Boulding, "The Economics of the Coming Spaceship Earth," in *Toward a Steady-State Economy*, edited by Herman E. Daly (San Francisco: W. H. Freeman, 1973), p. 127.
28 Ibid.
29 John S. Mill, *Principles of Political Economy*, vol. 2 (London: John W. Parker and Son, 1857), pp. 320-26, quoted in Daly, *Toward a Steady-State Economy*, p. 23.
30 Ibid., p. 13.
31 Daly, "Introduction," *Toward a Steady-State Economy*, p. 11.
32 Ibid., pp. 14-15.
33 Ibid., p. 23.
34 William Ophuls, *Ecology and the Politics of Scarcity* (San Francisco, W. H. Freeman, 1977), p. 154.
35 Garrett Hardin, "The Tragedy of the Commons," *Science* 162 (3 December 1968): 1243–48.
36 Ibid.
37 Robert L. Heilbroner, *Inquiry into the Human Prospect*, rev. ed. (New York: W. W. Norton, 1980), p. 130.
38 Ophuls, *Ecology and the Politics of Scarcity*, p. 152.
39 Ibid., p. 218.
40 E. F. Schumacher, *Small Is Beautiful: Economics as if People Mattered* (New York: Harper, 1973).
41 David W. Orr, "The Meadowcreek Project: A Model Sustainable Community," *Futurist* 15, 3 (June 1981): 47–53.
42 Thomas W. Foster, "Amish Society: A Relic of the Past Could Become a Model for the Future," *Futurist* 15, 8 (December 1981): 33–40.
43 *The Planetary Bargain: Proposals for a New Economic Order to Meet Human Needs* (Princeton, N.J.: Aspen Institute for Humanistic Studies, 1975).

44 Amilcar O. Herrera et al., *Catastrophe or a New Society? a Latin American World Model* (Ottawa: International Development Research Center, 1976).
45 McGeorge Bundy, "After the Deluge, the Covenant," *Saturday Review World* (24 August 1974), p. 18.
46 Peter Steinhart, "The Second Green Revolution," *New York Times Magazine* (25 October 1981), pp. 46–50.
47 Gerard K. O'Neill, *2081: A Hopeful View of the Human Future* (New York: Simon and Schuster, 1981), pp. 65–66.
48 Ibid., pp. 66–68.

The Environmental Issues Revisited

1 The discussion of the five factors is drawn heavily from Robin Attfield, *The Ethics of Environmental Concern* (New York: Columbia University Press, 1983), pp. 9–17.
2 Lester R. Brown et al., *State of the World 1984* (New York: W. W. Norton, 1984), p. 4.
3 Julian L. Simon, *The Ultimate Resource* (Princeton, N.J.: Princeton University Press, 1981), p. 9.
4 Ibid., p. 11.
5 See Parker G. Marden et al., *Population in the Global Arena* (New York: Holt, Rinehart and Winston, 1982), a text that uses the same framework as this volume, for a discussion of the role of population in other issues.
6 Barry Commoner, *The Closing Circle* (London: Jonathan Cape, 1982), p. 139; discussed in Attfield, *Ethics of Environmental Concern*, p. 12.
7 Ibid., p. 13.
8 The lists of negative and positive aspects of the Stockholm Conference are taken from Lynton K. Caldwell, *International Environmental Policy: Emergence and Dimensions* (Durham, N.C.: Duke University Press, 1984), p. 49.
9 Martin W. Holdgate, Mohammed Kassas, and Gilbert F. White, eds., *The World Environment 1972–1982: A Report by the United Nations Environment Programme* (Dublin: Tycooly International, 1982).
10 Ibid., p. 239.
11 Ibid., p. 351.
12 Leon F. Bouvier, "Planet Earth 1984–2034: A Demographic Vision," *Population Bulletin* 39, 1 (February 1984): 13.
13 Barnett F. Baron, "Population, Development, and Multinational Corporations," speech to the General Assembly of Legal Counsel, the Coca-Cola Company, San Diego, 12 December 1979.
14 United Nations, Department of International Economic and Social Affairs, *International Migration Policies and Programmes: A World Survey*, Population Studies No. 80 (New York: United Nations, 1982), p. 84.
15 J. P. Robinson, *The Effects of Weapons on Ecosystems*, United Nations Environment Programme Studies, vol. I (Elmsford, N.Y.: Pergamon Press, 1979), quoted in Holdgate, *World Environment 1972–1982*, p. 596.
16 Carl Sagan, "Nuclear War and Climatic Catastrophe: Some Policy Implications," *Foreign Affairs* 62 (Winter 1983/84): 259.
17 Ibid., pp. 263–64.
18 Ibid, p. 277.
19 L. Tangley, "U.N. Holds Global Environment Meeting," *Science News* 121

(29 May 1982): 358.

20 Patricia Scharlin, "Nairobi Remembers the Stockholm Conference," *Sierra* 67 (September/October 1982): 27.

21 Gilbert F. White, "Ten Years After Stockholm," *Science* 216 (7 May 1982): 569.

22 Office of Technology Assessment, U.S. Congress, *Global Models, World Futures, and Public Policy* (Washington, D.C.: U.S. Government Printing Office, 1982), pp. 3–4.

23 Kenneth E. Boulding, "The Idea of a Public," in *Managing the Commons*, edited by Garrett Hardin and John Baden (San Francisco: W. H. Freeman, 1977), p. 291.

24 Bouvier, "Planet Earth 1984–2034," p. 35.

Bibliography

Environment as a Global Issue

The following books introduce students to the global environment issue. Lester R. Brown's *The Twenty-Ninth Day* (New York: Norton, 1978) is an excellent introduction and overview of global environmental problems. It reviews not only the major ecosystems and the major problem areas (such as food, energy, and population), but also discusses problems of wealth distribution and the kinds of changes needed for mankind to survive. Lynton K. Caldwell's *Environment* (New York: Anchor Books, 1970) is an early general introduction to the challenges that environment presents to modern society and is still one of the best. Barry Commoner's *The Closing Circle* (New York: Knopf, 1971) is also written for the general public. This book explores the ways in which modern technologies, particularly nonbiodegradable ones, have greatly increased ecological problems. Harold and Margaret Sprout's *The Context of Environmental Politics* (Lexington: University Press of Kentucky, 1978), meant as a citizen's guide to the coming century, provides a balanced view of the dangers and possibilities facing us.

Environmental Quality (Washington, D.C.: Council on Environmental Quality), an annual report, is perhaps the best single collection of data and information on environmental problems. While largely focusing on the United States, it does include a chapter on the international environment. William Ophuls' *Ecology and the Politics of Scarcity* (San Francisco: Freeman, 1977), aimed at a college audience, gives an excellent overview of global environmental problems and trends. It also brings out the social and political challenges and changes needed for transition to a postindustrial society. Dennis Pirages' *Global Ecopolitics* (North Scituate, Mass.: Duxbury Press, 1978) is an international relations textbook written from the viewpoint of global environment and ecology. It places many traditional issues in new light. It also deals effectively with the important issues of technology and development. E. F. Schumacher's *Small Is Beautiful* (New York: Harper, 1973) is already a minor classic in the field of appropriate technology. This book also is concerned with basic environmental issues and ethics, especially as they relate to energy, economics, development, and education. *World Conservation Strategy* (New York: UNIPUB, 1980) is a report prepared by the International Union for the Conservation of Nature with support from several U.N. specialized agencies and is concerned with the rapid loss of global living and genetic resources. It outlines strategies for conserving and protecting them.

Actors

Three excellent introductions to the general subject of international organizations are Harold Jacobson's *Networks of Interdependence* (New York: Alfred A. Knopf, 1984, second edition); A. LeRoy Bennett's *International Organizations* (Englewood Cliffs, N.J.: Prentice-Hall, 1980, second edition); and *International Organizations: A Comparative Approach,* by Werner J. Feld and Robert S. Jordan (New York: Praeger, 1983). The Bennett book also presents a summary of the major global conferences of the 1970s.

The most comprehensive treatment of actors in the global environment, including UNEP's activities, can be found in Lynton Keith Caldwell's *International Environmental Policy: Emergence and Dimensions* (Durham, N.C.: Duke University Press, 1984).

Values

An especially influential book in the substantial literature on environmental perspectives and the philosophies underlying them is Paul Shepard and Daniel McKinley, eds., *The Subversive Science: Essays toward an Ecology of Man* (Boston: Houghton Mifflin, 1976). *Man's Responsibility for Nature: Ecological Problems and Western Traditions,* by John Passmore (New York: Scribner, 1974), is a good discussion of a variety of environmental values. For analyses of the philosophical underpinnings of environmentalism, see William T. Blackstone, ed., *Philosophy and Environmental Crisis* (Athens: University of Georgia Press, 1974). Two books that challenge humanistic orientations toward environmental values are Christopher D. Stone, *Should Trees Have Standing?* (Los Altos, Calif.: William Kaufmann, 1974), and Tom Regan and Peter Singer, eds., *Animal Rights and Human Obligations* (Englewood Cliffs, N.J.: Prentice-Hall, 1976). Concerning the prominence of the environment on national and international agendas, see Ervin Laszlo et al., *Goals for Mankind: A Report to the Club of Rome on New Horizons of Global Community* (New York: E. P. Dutton, 1977).

For further readings on the values of controlling pollution and preserving genetic diversity, see the policy section of this bibliography. One book on the consequences of pollution not mentioned there is Erik P. Eckholm, *The Picture of Health: Environmental Sources of Pollution* (New York: W. W. Norton, 1977). For a discussion of the conservation of resources and the entropy problem, see Jeremy Rifkin, *Entropy: A New World View* (New York: Viking Press, 1980). Books on related subjects include Dennis C. Pirages, ed., *The Sustainable Society: Implications for Limited Growth* (New York: Praeger, 1977), and Lester R. Brown, *Building a Sustainable Society* (New York: Norton, 1981). See also a special issue of the *International Studies Quarterly* (21, no. 4, December 1977) on the topic of "The Politics of Scarcity." A more extensive discussion of values related to population can be found in another volume in this series, *Population in the Global Arena,* by Parker G. Marden, Dennis C. Hodgson, and Terry L. McCoy (New York: Holt, Rinehart and Winston, 1982). The debate over "lifeboat" ethics was initiated by Garrett Hardin, "Living on a Lifeboat" (*Bioscience* 24, no. 10, October 1974, pp. 561–68). He elaborates on the subject in *The Limits to Altruism: An Ecologist's View of Survival* (Bloomington, Ind.: Indiana University Press, 1977). Among the many responses to Hardin's argument are Peter G. Brown, ed., *Food Policy: The Responsibility of the United States on Life and Death Choices* (New York: Free Press, 1977), and a series of articles appearing in *Alternatives* (7, no. 3, Winter 1981–82).

Turning to the relationship between peace and environmental values, the reader will find conceptual discussions of the meaning of peace in Johan Galtung, "Violence, Peace, and Peace Research" (*Journal of Peace Research* 6, no. 3, 1969, pp. 167–91), and Marvin S. Soroos, "Adding an Intergenerational Dimension to Conceptions of Peace" (*Journal of Peace Research* 13, no. 13, 1976, pp. 173–83). A number of studies assess the impact of war on the environment in the context of Vietnam, including those by John Lewallen, *Ecology of Devastation: Indochina* (Baltimore: Penguin, 1971), and Raphael Littauer and Norman Uphoff, eds., *The Air War in Indochina* (Boston: Beacon Press, 1971, revised edition). *The United States and the Global Struggle for Minerals*, by Alfred E. Eckes, Jr. (Austin: University of Texas Press, 1979), contains an extensive historical analysis of the relationship between resource insecurities and war. Notable among many publications that discuss future security from the perspective of reliable access to resources are Lester R. Brown, *Redefining National Security*, Worldwatch Paper no. 14 (Washington, D.C.: Worldwatch Institute, 1977); David A. Deese, ed., *Energy and Security* (Cambridge, Mass.: Ballinger, 1981); and Daniel Yergen and Martin Hillenbrand, eds., *Global Insecurity: A Strategy for Energy and Economic Renewal* (Boston: Houghton Mifflin, 1982).

On the subject of economic development and the environment, there is an extensive literature on the "new international economic order" proposed by the Third World, notable examples of which are Jagdish Bhagwati, ed., *The New International Economic Order: The North-South Debate* (Cambridge, Mass.: MIT Press, 1977); Roger Hansen, *Beyond the North-South Stalemate* (New York: McGraw-Hill, 1979); and *North-South, a Program for Survival, a Report of the Independent Study Commission on International Development Issues under the Chairmanship of Willy Brandt* (Cambridge, Mass.: MIT Press, 1980). For background readings on the initial reservations of the less developed countries to international environmental action, see the Founex Report, "Environment and Development" (*International Conciliation*, no. 568, January 1972). Several studies have attempted to demonstrate that natural resources are sufficient to sustain the economic development of all countries, among which are the United Nations study by Wassily Leontieff, *The Future of the World Economy* (New York: Oxford University Press, 1976); Jan Tinbergen, Antony J. Dolman, and Jan van Ettinger, eds., *Reshaping the International Order: A Report to the Club of Rome* (New York: E. P. Dutton, 1976); and Antony J. Dolman, *Resources, Regimes, World Order* (New York: Pergamon Press, 1980). An Indian scholar, Rajni Kothari, "Environment and Alternative Development" (*Alternatives* 5, no. 4, January 1980, pp. 427–75), suggests ways of redefining development in the context of environmental values.

Policy

Garrett Hardin's essay "The Tragedy of the Commons" appeared originally in *Science* (vol. 162, December 13, 1968, pp. 1243–48). He elaborates his theories more extensively in *Exploring New Ethics for Survival* (Baltimore: Penguin, 1972), chapter 13 of which is an especially useful discussion of strategies for averting the tragedy. A book Hardin edited with John Baden, *Managing the Commons* (San Francisco: W. H. Freeman, 1977), contains a variety of articles that discuss problems pertaining to the management of common resources. Marvin S. Soroos, "The Commons and Lifeboat as Guides for International Ecological Policy" (*International Studies Quarterly* 21, no. 4, December 1977, pp. 647–74), applies Hardin's theories to the international sphere, as does Per Magnus Wijkman, "Managing the Global Commons" (*Interna-

tional Organization 36, no. 3, Summer 1982, pp. 511–36).

Two books that contain excellent discussions of the management of ocean fisheries are H. Gary Knight, *Managing the Sea's Living Resources: Legal and Political Aspects of High Seas Fisheries* (Lexington, Mass.: Lexington Books, 1977), and James Joseph and Joseph Greenough, *International Management of Tuna, Porpoise, and Billfish* (Seattle: University of Washington Press, 1979). Robert Mandel, "Transnational Resource Conflict: the Politics of Whaling" (*International Studies Quarterly* 24, no. 1, March 1980, pp. 99–127), looks at the problem of whale conservation as a problem of the commons. For a more thorough treatment of the whale issue, see K. Radway Allen, *Conservation and Management of Whales* (Seattle: University of Washington Press, 1980), and Robert McNally, *So Remorseless a Havoc: Of Dolphins, Whales, and Men* (Boston: Little, Brown and Co., 1981). *The Whale Report* is a newsletter published four times annually by the Center for Environmental Education (624 9th St., N.W., Washington, D.C. 20001).

A number of books discuss the Law of the Sea issues in a more comprehensive way: two of the best are Ross D. Eckert, *The Enclosure of Natural Resources: Economics and the Law of the Sea* (Stanford, Calif.: Hoover Institution Press, 1979), and Ann L. Hollick, *U.S. Foreign Policy and the Law of the Seas* (Princeton, N.J.: Princeton University Press, 1981). Ocean issues are the subject of entire issues of the journals *International Organization* (vol. 31, no. 2, Spring 1977) and the *Columbia Journal of World Business* (vol. 15, no. 4, Winter 1980), the latter of which contains two articles written from a Third World perspective.

Several books have been published recently on the preservation of species, the most noteworthy being Norman Myers, *The Sinking Ark: A New Look at the Problem of Disappearing Species* (New York: Pergamon, 1979), and Paul and Ann Ehrlich, *Extinction: The Causes and Consequences of the Disappearance of Species* (New York: Random House, 1981). The Myers book is of particular interest for its argument that species should be considered the common property of humanity. Robert Boardman's *International Organization and the Conservation of Nature* (Bloomington: Indiana University Press, 1981) gives a detailed history of the role of international organizations, both governmental and nongovernmental, in the preservation of species. A wealth of factual information on the CITES treaty and its implementation is contained in Tim Inskipp and Sue Wells, *International Trade in Wildlife* (London: Earthscan, 1979). For a fascinating history of the extinction of species in non-Western areas, see Robert M. McClung, *Lost Wild Worlds: the Story of Extinction and Vanishing Wildlife of the Eastern World* (New York: William Morrow, 1976).

The best single work on the problem of oil pollution and the international law that addresses it is by R. Michael M'Gonigle and Mark W. Zacher, *Pollution, Politics, and International Law: Tankers at Sea* (Berkeley and Los Angeles: University of California Press, 1979). A recent and definitive study of IMCO and the regulation of supertankers is Harvey B. Silverstein's *Superships and Nation-States: The Transnational Politics of the Intergovernmental Maritime Consultative Organization* (Boulder, Colo.: Westview Press, 1978). Noel Mostert's *Supership* (New York: Warner Books, 1974) is a readable best-seller describing the author's personal experience on board a supertanker; it presents much of the lore of these ships, especially the dangers they encounter. For briefer presentations of existing ocean law, see Bernard J. Abrahamsson, "The Maritime Environment and Ocean Shipping" (*International Organization* 31, no. 2, Spring 1977, pp. 281–322), and a two-part article appearing in the journal *Oceans* by Elliot L. Richardson, who represented the Carter administration at the Law of the Sea talks, entitled "Prevention of Vessel-Source Pollution" (vol. 13, no. 2, March 1980, pp. 2–5, and vol. 13, no. 4, 1980, pp. 58–61).

A good analysis of general international public policy on pollution is Jan Schneider, *World Public Order of the Environment: Towards an International Ecological Law and Organization* (Toronto: University of Toronto Press, 1979). On the same subject, see an earlier collection of essays edited by Ludwik A. Teclaff and Albert E. Utton, *International Environmental Law of Pollution* (New York: Praeger, 1974). James Barros and Douglas M. Johnston provide a thorough review of international case law in *The International Law of Pollution* (New York: Free Press, 1974).

On the subject of acid rain, the reader is referred to Gene E. Likens et al., "Acid Rain" (*Scientific American* 241, no. 4, October 1979, pp. 39–47), and to Anne LaBastille, "Acid Rain: How Great a Menace?" (*National Geographic* 160, no. 5, November 1981, pp. 652–81). The ECE treaty on transboundary air pollution is described by Armin Rosencranz in "The Problem of Transboundary Pollution" (*Environment* 22, no. 15, June 4, 1980, pp. 15–20).

Future

A vast literature is potentially relevant to global environmental futures. Two journals useful for keeping informed about the field are *The Futurist* and *Futures: A Journal of Forecasting and Planning.* For methods of studying the future, see a collection of articles edited by Louis R. Beres and Harry R. Targ, *Planning Alternative World Futures: Values, Methods, and Models* (New York: Praeger, 1975). For a summary of many of the works cited in this section, see chapter 2 of Antony J. Dolman, *Resources, Regimes, World Order* (New York: Pergamon, 1981).

One of the earlier forecasts of future environmental problems was a book by Harrison S. Brown, *The Challenge of Man's Future* (New York: Viking, 1954). Basic reading on the subject also includes the "doomsday" reports briefly summarized in the text: Donella Meadows et al., *The Limits to Growth* (New York: Universe, 1972); Mihajlo Mesarovic and Eduard Pestel, *Mankind at the Turning Point* (New York: E. P. Dutton, 1974); and *The Global 2000 Report to the President: Entering the Twenty-First Century* (New York: Penguin Books, 1980), prepared by the U.S. Council on Environmental Quality and the Department of State. Among the more noteworthy critiques of these studies are H. S. D. Cole et al., *Models of Doom: A Critique of the Limits to Growth* (New York: Universe, 1973); Wilfred Beckerman, *Two Cheers for the Affluent Society: A Spirited Defense of Limited Growth* (New York: St. Martin's Press, 1974); Herman Kahn et al., *The Next Two Hundred Years: A Scenario of America and the World* (New York: William Morrow, 1976); and Julian L. Simon, *The Ultimate Resource* (Princeton, N.J.: Princeton University Press, 1981). *Groping in the Dark,* by Donella Meadows et al. (New York: Wiley, 1982), reflects on the first decade of global modeling, as does a special issue of the journal *Futures* on the topic "Global Modeling Revisited."

Turning to the subject of preferred futures, the reader might begin with Saul H. Mendlovitz, ed., *On the Creation of a Just World Order: Preferred Futures for the 1990s* (New York: Free Press, 1975), for summary articles of the WOMP designs of the Institute for World Order. The Institute also publishes *Alternatives: A Journal of World Policy,* which contains scholarly articles on a number of the subjects discussed in this chapter. Two seminal books on the subject of functionalism are those by David Mitrany, *A Working Peace System* (Chicago: Quadrangle, 1966), and Ernst B. Haas, *Beyond the Nation State: Functionalism and International Organization* (Stanford, Calif: Stanford University Press, 1964). Leading works on the subject of international regimes and policies include Evan Luard, *International Agencies: the Emerging Framework of Interdependence* (Dobbs Ferry, N.Y.: Oceana, 1977); Robert O. Keohane and Joseph Nye, *Power and Independence: World Politics in Transition* (Boston: Little, Brown and Co., 1977); and Harold K. Jacobson, *Networks of Interdependence: International Organizations and the Global Political System* (New York: Alfred A. Knopf, 1979). The topic is nicely summarized in an article by Oran R. Young, "International Regimes: Problems of Concept Formation" (*World Politics* 32, no. 3, April 1980, pp. 331–56). A recent contribution to the field is Gary K. Bertsch, ed., *Global Policy Studies* (Beverly Hills, Calif.: Sage Publications, 1982). The journal *International Organization* includes numerous articles on international regimes and policies.

Highly recommended on the topic of steady-state economics are two collections of essays edited by Herman Daly, which have some overlapping content: *Toward a Steady-State Economy* (San Francisco: W. H. Freeman, 1973), and *Economics, Ecology, Ethics: Essays Toward a Steady-State Economy* (San Francisco: W. H. Freeman, 1980). Other books on the subject include Mancur Olson and Hans H. Landsberg, eds., *The No-Growth Economy* (New York: W. W. Norton, 1973), and Edward F. Renshaw, *The End of Progress: Adjusting to a No-Growth Economy* (North Scituate, Mass.: Duxbury Press, 1976). Amory B. Lovins is well known for his writings on renewable energy sources; see "Energy Strategy: the Road Not Taken" (*Foreign Affairs* 55, 1976, pp. 65–96), and *Soft Energy Paths* (Cambridge, Mass.: Ballinger, 1977). For practical tips on steady-state living, see the magazine *The Mother Earth News.* An interesting novel that portrays a steady-state lifestyle is *Ecotopia* by Ernest Callenbach (New York: Bantam Books, 1975).

The most noteworthy advocates of the need for strong, centralized power to avoid environmental catastrophe are those by Robert L. Heilbroner, *An Inquiry into the Human Prospect: Updated and Reconsidered for the 1980s* (New York: W. W. Norton, 1980), and William Ophuls, *Ecology and the Politics of Scarcity* (San Francisco: W. H. Freeman, 1977). See in particular chapter 4 of the Ophuls book, which summarizes some of the relevant theories of Hobbes, Locke, and Smith. E. F. Schumacher has described the alternative to decentralized systems in *Small Is Beautiful: Economics as if People Mattered* (New York: Harper, 1973). Murray Bookchin has written several books on the subject, the most recent being *The Colony of Freedom* (Palo Alto, Calif.: Cheshire Books, 1982). A similar theme pertaining to food can be found in Frances Moore Lappe and Joseph Collins, *Food First: Beyond the Myth of Food Scarcity* (Boston: Houghton Mifflin, 1977). See also Bruce Stokes, *Helping Ourselves: Local Solutions to Global Problems* (New York: W. W. Norton, 1981).

Future world orders that emphasize global equity taking into account natural limits are set forth by Richard A. Falk, *This Endangered Planet: Prospects and Proposals for Human Survival* (New York: Random House, 1972); Jan Tinbergen, *Reshaping the International Order: A Report to the Club of Rome* (New York: E. P. Dutton, 1976); Harlan Cleveland, *The Third Try at World Order: U.S. Policy for an Interdependent World* (New York: Aspen Institute, 1976); and Wassily Leontief et al., *The Future of the World Economy* (New York: Oxford University Press, 1977). Since 1974 the Overseas Development Council has annually published a book on economic development, food, and resource issues entitled *U.S. Foreign Policy and the Third World,* the series title being *Agenda (Year)* (New York, Praeger). For an excellent analysis of the future of North-South relationships, see Roger D. Hansen, *Beyond the North-South Stalemate* (New York: McGraw-Hill, 1979), one of the books in the 1980s Project of the Council on Foreign Relations.

The possibilities and potential problems with high technology solutions to environmental problems have been widely discussed. On the subject of recombinant DNA, see Richard Hutton, *Bio-Revolution: DNA and the Ethics of Man-Made Life* (New York: Mentor, 1978); David A. Jackson and Stephen P. Stich, eds., *The Recombinant DNA Debate* (Englewood Cliffs, N.J.: Prentice-Hall, 1979); and Clifford Brobstein, *The Image of the Double Helix: The Recombinant DNA Debate* (San Francisco: W. H. Freeman, 1979). The possibilities for space colonies are discussed by Gerard K. O'Neill in his books *The High Frontier* (New York: William Morrow, 1977) and *2081: A Hopeful View of the Human Future* (New York: Simon and Schuster, 1981). For critiques of the space colony idea, see Jack D. Salmon, "Politics of Scarcity versus Technological Optimism: A Possible Reconciliation?" (*International Studies Quarterly* 21, no. 4, December 1977, pp. 701–20), and Daniel Deudney, "Space Industrialization: The Mirage of Abundance" (*Futurist* 16, no. 6, December 1982, pp. 47–53).

The Environment Issue Revisited

The most comprehensive treatment of the state of the contemporary environment is *The World Environment 1972–1982: A Report by the United Nations Environment Programme* (Dublin: Tycooly International Publishing, 1982). Edited by Martin W. Holdgate, Mohammed Kassas, and Gilbert F. White, this report was published for the United Nations Environment Program (UNEP). UNEP publishes an annual report on the state of the environment. The coverage is not comprehensive; rather, it focuses on selected topics, which are typically titled *The State of the Environment—Selected Topics.* Finally, for an analysis of how the plan developed at Stockholm has been implemented, see UNEP's *Review of the Implementation of the Stockholm Action Plan* (Nairobi, 1982).

"Acid Precipitation." *Mother Earth News* 73 (January/February 1982): 123.

Attfield, Robin. *The Ethics of Environmental Concern.* New York: Columbia University Press, 1983.

Baron, Barnett F. "Population, Development, and Multinational Corporations." Speech to General Assembly of Legal Counsel, the Coca-Cola Company, 12 December 1979, San Diego. Mimeographed.

Barros, James, and Johnston, Douglas M. *The International Law of Pollution.* New York: Free Press, 1974.

Beckerman, Wilfred. *Two Cheers for the Affluent Society: A Spirited Defense of Economic Growth.* New York: St. Martin's Press, 1974.

Bennett, A. LeRoy. *International Organizations.* Englewood Cliffs, N.J.: Prentice-Hall, 1980.

Boardman, Robert. *International Organization and the Conservation of Nature.* Bloomington: Indiana University Press, 1981.

Boulding, Kenneth D. "The Economics of the Coming Spaceship Earth." In *Toward a Steady-State Economy,* edited by Herman E. Daly. San Francisco: W. H. Freeman, 1973.

———. "The Idea of a Public." In *Managing the Commons,* edited by Garrett Hardin and John Baden. San Francisco: W. H. Freeman, 1977.

Bouvier, Leon F. "Planet Earth 1984–2034: A Demographic Vision." *Population Bulletin* 39, 1 (February 1984).

Brown, Joseph E. *Oil Spills: Danger in the Sea.* New York: Dodd Mead, 1978.

Brown, Lester R. *Redefining National Security*. Worldwatch Paper 14. Washington, D.C.: Worldwatch Institute, 1977.
———. *The Twenty-Ninth Day*. New York: W. W. Norton, 1978.
———, et al. *State of the World 1984*. New York: W. W. Norton, 1984.
Bundy, McGeorge. "After the Deluge, the Covenant." *Saturday Review World* (24 August 1974): 18–20.
Caldwell, Lynton K. *International Environmental Policy: Emergence and Dimensions*. Durham, N.C.: Duke University Press, 1984.
Cleveland, Harlan. *The Third Try at World Order*. Princeton, N.J.: Aspen Institute for Humanistic Studies, 1976.
Council on Environmental Quality. *Environmental Quality—1979*. Washington, D.C.: U.S. Government Printing Office, 1980.
——— and the Department of State. *The Global 2000 Report to the President. Volume I*. Washington, D.C.: U.S. Government Printing Office, 1980.
———. *The Global 2000 Report to the President. Volume II: The Technical Report*. Washington, D.C.: U.S. Government Printing Office, 1980.
Daly, Herman E., ed. *Toward a Steady-State Economy*. San Francisco: W. H. Freeman, 1973.
Eckert, Ross D. *The Enclosure of Ocean Resources: Economics and the Law of the Sea*. Stanford, Calif.: Hoover Institute Press, 1979.
Eckes, Alfred E., Jr. *The United States and the Global Struggle for Minerals*. Austin: University of Texas Press, 1979.
Eckholm Erik P. *Disappearing Species: The Social Challenge*. Washington, D.C. Worldwatch Institute, July 1978.
———. *Down to Earth: Environment and Human Needs*. New York: W. W. Norton, 1982.
———. *The Picture of Health: Environmental Sources of Disease*. New York: W. W. Norton, 1977.
The Effects of Nuclear War. Washington, D.C.: U.S. Arms Control and Disarmament Agency, 1979.
Ehrenfeld, David. *The Arrogance of Humanism*. New York: Oxford University Press, 1978.
Ehrlich, Paul, and Ehrlich, Ann. *Extinction: the Causes and Consequences of the Disappearance of Species*. New York: Random House, 1981.
Ehrlich, Paul R., Ehrlich, Ann H., and Holdren, J. P. *Human Ecology: Problems and Solutions*. San Francisco: W. H. Freeman, 1973.
Feraru, Anne Thompson. "Transnational Political Interests and the Global Environment." *International Organization* 28 (Winter 1974): 310–36.
Foster, Thomas W. "Amish Society: A Relic of the Past Could Become a Model for the Future." *Futurist* 15, 8 (December 1981): 33–40.
Gilman, Katherine. "Julian Simon's Cracked Crystal Ball." *The Public Interest* 65 (Fall 1981): 74–77.
Greenwood, N. J., and Edwards, J. M. B. *Human Environments and Natural Systems*. 2d ed. North Scituate, Mass.: Duxbury Press, 1979.
Group of Experts on the Scientific Aspects of Marine Pollution. *Impact of Oil on the Marine Environment*. Rome: Food and Agriculture Organization, 1977.
Grove, Noel. "Wild Cargo: The Business of Smuggling Animals." *National Geographic* 159, 3 (March 1981): 287–315.
Gulland, J. A. "Open Ocean Resources." In *Fisheries Management*, edited by Robert T. Lackey and Larry A. Neilsen. New York: John Wiley, 1980.

Hardin, Garrett. "Living on a Lifeboat." *Psychology Today* 24, 10 (October 1974): 563–65.
———. *The Limits to Altruism: An Ecologist's View of Survival*. Bloomington: Indiana University Press, 1977.
———. "The Tragedy of the Commons." *Science* 162 (3 December 1968): 1243–48.
Heilbroner, Robert L. *Inquiry into the Human Prospect*. Revised Edition. New York: W. W. Norton, 1980.
Herrera, Amilcar O., et al. *Catastrophe or a New Society? a Latin American Model*. Ottawa: International Development Research Center, 1976.
Holdgate, Martin W., Kassas, Mohammed, and White, Gilbert F., eds. *The World Environment 1972–1982: A Report by the United Nations Environment Programme*. Dublin: Tycooly International Publishing, 1982.
James, Gregory. "Yemeni Wealth Decimates Rhinos." *New York Times*, 9 March 1982.
Juda, Lawrence. "International Environmental Concern: Perspectives of and Implications for Developing States." In *The Global Predicament: Ecological Perspectives on World Order*, edited by David W. Orr and Marvin S. Soroos. Chapel Hill: University of North Carolina Press, 1979.
Kahn, Herman, Brown, William, and Martel, Leon. *The Next 200 Years: A Scenario for America and the World*. New York: William Morrow, 1976.
Kaufman, Michael T. "Preserving Rare Species is an Ironic Success." *New York Times*, 8 March 1981, IV, 20.
Kiechell, Walter III. "Admiralty Case of the Century," *Fortune* 99 (23 April 1979): 78–82.
Knight, H. Gary. *Managing the Sea's Living Resources: Legal and Political Aspects of High Seas Fisheries*. Lexington, Mass.: Lexington Books, 1977.
Knight, Jeffrey. "Clean Air III: The Global Dimension." *Sierra Club Bulletin* 66, 3 (May/June 1981): 15–18.
LaBastille, Anne. "Acid Rain: How Great the Threat?" *National Geographic* 160, 5 (November 1981).
Likens, Gene E., et al. "Acid Rain." *Scientific American* 241, 4 (October 1979): 39–47.
M'Gonigle, R. Michael, and Zacher, Mark W. *Pollution, Politics, and International Law: Tankers at Sea*. Berkeley: University of California Press, 1979.
Marden, Parker G., et al. *Population in the Global Arena*. New York: Holt, Rinehart and Winston, 1982.
Meadows, Donella H., et al. *The Limits to Growth*. New York: Universe Books, 1971.
Mesarovic, Mihajlo, and Pestel, Edward. *Mankind at the Turning Point*. New York: E. P. Dutton, 1974.
Mineral Commodity Summaries. Washington: U.S. Bureau of Mines, 1981.
Murray, Geoffrey. "Japan, Angered by World Whaling Ban, Gets Ready to Defy It." *Christian Science Monitor*, 27 July 1982.
Myers, Norman. *The Sinking Ark: A New Look at the Problem of Disappearing Species*. Elmsford, N.Y.: Pergamon Press, 1979.
Office of Technology Assessment, U.S. Congress. *Global Models, World Futures, and Public Policy*. Washington, D.C.: U.S. Government Printing Office, 1982.
———. *The Effects of Nuclear War*. Washington, D.C.: U.S. Government Printing Office, 1979.
O'Neill, Gerard K. *2081: A Hopeful View of the Human Future*. New York: Simon and Schuster, 1981.
Ophuls, William. *Ecology and the Politics of Scarcity*. San Francisco: W. H.

Freeman, 1977.
Orr, David W. "The Meadowcreek Project: A Model Sustainable Community." *Futurist* 15, 3 (June 1981): 47–53.
Payne, Millie. "Status of Whales." *The Living Wilderness* 43, 147 (December 1979): 16–17.
Petroleum in the Marine Environment. Washington, D.C.: Academy of Sciences, 1975.
The Planetary Bargain: Proposals for a New Economic Order to Meet Human Needs. Princeton, N.J.: Aspen Institute for Humanistic Studies, 1975.
Prosterman, Roy L. *Surviving to 3000: An Introduction to the Study of Lethal Conflict*. Belmont, Calif.: Wadsworth, 1972.
Rappoport, Roy A. *Pigs for the Ancestors: Ritual in the Ecology of a New Guinea People*. New Haven: Yale University Press, 1967.
Regenstein, Lewis. *The Politics of Extinction: The Shocking Story of the World's Endangered Species*. New York: Macmillan, 1975.
Rifkin, Jerry. *Entropy: A New World View*. New York: Viking Press, 1980.
Robinson, J. P. *The Effects of Weapons on Ecosystems*. United Nations Environment Programme Studies, Vol. I. New York: Oxford University Press, 1979.
Sagan, Carl. "Nuclear War and Climatic Catastrophe: Some Policy Implications." *Foreign Affairs* 62 (Winter 1983/84): 256–92.
Scharlin, Patricia. "Nairobi Remembers the Stockholm Conference." *Sierra* 67 (September–October 1982): 26–28.
Schumacher, E. F. *Small Is Beautiful: Economics as if People Mattered*. New York: Harper, 1973.
Simon, Julian L. *The Ultimate Resource*. Princeton, N.J.: Princeton University Press, 1981.
Sivard, Ruth Leger. *World Military and Social Expenditures—1982*. New York: Institute for World Order, 1982.
Smith, W. Eugene. *Minamata*. New York: Holt, Rinehart and Winston, 1975.
Soroos, Marvin S. "Exploring Global Ecological Futures." In *The Global Predicament: Ecological Perspectives on World Order*, edited by David W. Orr. Chapel Hill: The University of North Carolina Press, 1979.
———. "Trends in the Perception of Ecological Problems in the United Nations General Debates." *Human Ecology* 9, 1 (1981): 32–35.
Steinhart, Peter. "The Second Green Revolution." *New York Times Magazine*, 25 October 1981, pp. 46–50.
Tangley, L. "U.N. Holds Global Environment Meeting." *Science News* 121 (29 May 1982).
Thant, U. "Ten Critical Years." *UN Monthly Chronicle* 6, 7 (July 1969): ii.
United Nations. *Declaration Adopted by the UNEP Session of a Special Character—Nairobi, 18 May 1982*. Nairobi, Kenya: UNEP Liaison Office, 1982.
———. *Report of the United Nations Conference on the Human Environment, Stockholm, 5–16 June 1972*. UN Document A/Conf. 48/14/Rev. 1. New York, 1973.
———. Department of International and Social Affairs. *International Migration Policies and Programmes: A World Survey*. Population Studies No. 80. New York: United Nations, 1982.
"UN Conference on New and Renewable Sources of Energy." *U.S. Department of State Bulletin* (January 1982): 63–78.
The United Nations Environment Program. Nairobi, Kenya: UNEP, 1979.

U.S. Department of Agriculture and the Council on Environmental Quality. *National Agricultural Lands Study: Final Report 1981*. Washington, D.C.: U.S. Government Printing Office, 1981.

Wagner, Richard W. *Environment and Man*. 3d ed. New York: W. W. Norton, 1978.

Webster, Bayard. "Tanzania Asks Rich Countries to Aid Wildlife." *New York Times*, 29 March 1979.

Westing, Arthur H. "Indochina: Prototype of Ecocide." In *Air, Water, Earth, Fire: The Impact of the Military on World Environmental Order*, by Michael McClintock et al., pp. 16–23. San Francisco: Sierra Club, 1974.

White, Gilbert F. "Ten Years After Stockholm," *Science* 216 (7 May 1982): 569.

"World Abortion Trends." *Population* 9 (April 1979): 1.

Index

Contributors

Kenneth A. Dahlberg is Professor of Political Science at Western Michigan University. He has published extensively in the field of global environmental issues, and his book *Beyond the Green Revolution* won the 1981 Sprout and Sprout Award of the International Studies Association (ISA) for the best book on international environmental affairs. He helped establish and was the first Chairperson of the Environmental Studies Section of the ISA. He is a Fellow of the American Association for the Advancement of Science (AAAS). From 1980 to 1982 he served as a member of the U.S. National Committee for Man and the Biosphere. He is currently directing an interdisciplinary team of fourteen scholars, under a grant from the National Science Foundation and the National Endowment for the Humanities, to outline the new directions in agricultural research that are needed to take account of a number of neglected resource, social, ethical, and environmental issues.

Marvin S. Soroos is currently Associate Professor of Political Science and Public Administration at North Carolina State University in Raleigh, where he has taught graduate and undergraduate courses on global issues and social science methodology since 1970, including a course entitled Global Environmental Politics. With David Orr, he edited *The Global Predicament: Ecological Perspectives on World Order.* His published articles have appeared in *International Organization, International Studies Quarterly, Human Ecology,* and *The Journal of Peace Research.* Among his present projects is a book that compares international policy responses to six problems having global dimensions: nuclear proliferation, economic development, human rights, pollution, ocean resources, and telecommunications. Since 1981 he has served as the Chairman of the Environmental Studies Section of the International Studies Association.

Anne Thompson Feraru is Professor of Political Science at California State University, Fullerton, and previously taught at the University of Hawaii, Hunter College, and Wilson College. Her research on international nongovernmental organizations has been published in *International Organization, Transnational Associations,* and in William Evan, ed., *Knowledge and Power in a Global Society.* She was an NGO observer at the 1976 UN Conference on Human Settlements and the 1979 UN Conference on Science and Technology for Development, and in 1980 served as a volunteer staff member at the Environment Liaison Center in Nairobi.

James E. Harf is Professor of Political Science and Mershon Senior Faculty at The Ohio State University. He was a Visiting Professor at Duke University in 1978–79 and an American Council on Education Fellow in Academic Administration at Pennsylvania State University in 1975–76. He served for six months as the principal undergraduate consultant on President Carter's Foreign Language and International Studies Commission. He has been the recipient of twenty grants that have focused primarily on materials and faculty development. He is currently codirector (with B. Thomas Trout) of a U.S. Department of Education grant to develop international materials to be included in five types of courses in the general business curriculum. He also serves with Trout as codirector of a project funded by the Ford Foundation to introduce national security concepts and topics into five standard secondary social studies courses. He is the author and editor of many books in international relations, including *National Security Affairs: Theoretical Perspectives and Contemporary Issues* and *International and Comparative Politics: A Handbook.* He is Executive Director of the Consortium for International Studies Education.

B. Thomas Trout is Associate Professor of Political Science at the University of New Hampshire. He is Chairman of the Consortium for International Studies Education and a member of the Executive Committee of the International Studies Association. He is co-editor and author of the recently published text *National Security Affairs: Theoretical Perspectives and Contemporary Issues* (1982) and has published articles in the *American Political Science Review, International Studies Quarterly, Naval War College Review,* and other journals.